Mark Bomberg, David W. Yarbrough, Hamed H. Saber
**Retrofitting**

# Also of interest

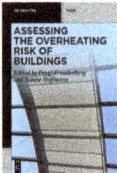

*Assessing the Overheating Risk of Buildings*
Peggy Freudenberg, Sabine Hoffmann (Eds.), 2024
Open Access Publication
ISBN 978-3-11-131802-8, e-ISBN (PDF) 978-3-11-131865-3,
e-ISBN (EPUB) 978-3-11-131968-1

*Designing with Multi-Agent Systems.*
*A Computational Methodology for Form-Finding Using Behaviors*
Evangelos Pantazis, 2024
ISBN 978-3-11-079704-6, e-ISBN (PDF) 978-3-11-079743-5,
e-ISBN (EPUB) 978-3-11-079747-3

*CAD/CAM.*
*Computer-Aided Design and Manufacturing*
Panagiotis Kyratsis, Athanasios Manavis, J. Paulo Davim, planned
publication: 2025
ISBN 978-3-11-158467-6, e-ISBN (PDF) 978-3-11-158506-2,
e-ISBN (EPUB) 978-3-11-158555-0

Mark Bomberg, David W. Yarbrough,
Hamed H. Saber

# Retrofitting

The Energy and Environment of Buildings

**DE GRUYTER**

**Authors**
Prof. Dr. Mark Bomberg
Mechanical and Aeronautical Dept.
Clarkson University, Potsdam, NY., USA
Innovative Building Systems
Radom, 26-611, Poland
mark.bomberg01@gmail.com

Prof. Dr. David W. Yarbrough, PhD, PE
R&D Services, Inc.
Watertown, Tennessee, USA 37184
dave@rdservices.com

Prof. Dr. Hamed H. Saber
Jubail Industrial College
Royal Commission of Jubail and Yanbu
Al Jubail 35718
Saudi Arabia
hhsaber@gmail.com

ISBN 978-3-11-914441-4
e-ISBN (PDF) 978-3-11-221702-3
e-ISBN (EPUB) 978-3-11-221751-1

**Library of Congress Control Number: 2025943860**

**Bibliographic information published by the Deutsche Nationalbibliothek**
The Deutsche Nationalbibliothek lists this publication in the Deutsche Nationalbibliografie;
detailed bibliographic data are available on the Internet at http://dnb.dnb.de.

© 2026 Walter de Gruyter GmbH, Berlin/Boston, Genthiner Straße 13, 10785 Berlin
Cover image: John_Lamb/E+/Getty Images
Typesetting: Integra Software Services Pvt. Ltd.
Printing and binding: CPI books GmbH, Leck

www.degruyterbrill.com
Questions about General Product Safety Regulation:
productsafety@degruyterbrill.com

# Contents

**Chapter 3**
**Air transport — 35**

**Chapter 4**
**Interacting water and vapor flows — 56**

# Executive summary in German
# Zielsetzung und Zusammenfassung des Buches

Eine Überprüfung der Passivhausanwendungen und Trends in der Nachrüstungstechnologie führten zu einem Mix aus klimamodifizierten Anwendungen, die mit Blick auf eine universelle Technologie ausgewählt wurden.

**Zielsetzungen**: Die heutige Bauphysikgeneration unterscheidet sich von der des 20. Jahrhunderts. Das liegt an einem Integrierten Designprozess (IDP), der eine individuelle Vision für das geplante Gebäude schafft, und an der Einbeziehung von Energiemodellierern. Unsere Generation beschäftigt sich bereits mit dem „Gebäude als System". Nun ist es an der Zeit, das Systemdenken auf neue Dimensionen auszuweiten. Das folgende Buch beschreibt einen neuen Ansatz, der als passive und thermoaktive Cluster-Systemtechnologie bezeichnet wird. Das Ziel dieses Buches ist es demnach, *eine Reihe von Methoden zu* entwickeln und zu diskutieren, *um neue oder nachgerüstete, nachhaltige Gebäude mit Null-Kohlenstoffemission und nahezu Null-Energieverbrauch zu bauen.* Wir gehen zudem davon aus, dass das Ziel der Bauphysik darin besteht, *verschiedene Methoden zu analysieren und zu vergleichen, die die gleichen Ziele verfolgen.*

Das Buch identifiziert die derzeit beste Technologie, und gleichzeitig müssen wir eine soziale Bewegung auslösen, um diese Technologie populär zu machen und ihre Erschwinglichkeit zu erhöhen. Zu diesem Zweck wird im vorliegenden Buch die Gründung eines sozioökonomischen Konsortiums vorgeschlagen, ähnlich denen, die in den 1980er und 1990er Jahren in Nordamerika aufgebaut wurden, jedoch mit dem Ziel, die Nachrüstungstechnologie mit der Verlangsamung des Klimawandels zu verbinden.

**Hintergrund:** Seit der BEST1-Konferenz (Bomberg and Onysko, 2008) befasst sich ein kleines, virtuelles Team von Ingenieuren mit der Herausforderung der Integration von Wärme-, Luft- und Wasserübertragungen mit *Überwachung und Modellierung, um die Leistung neuer Technologien für nachgerüstete Wohnungen und Gebäude* zu evaluieren und zu verbessern. Die Herausforderung besteht darin, aus der heutigen Technologie sowohl die kritischen Trends für die Zukunft als auch die neuen Testmethoden für integrierte Gebäude herauszuarbeiten. Die in diesem Buch vorgestellte Technologie *kombiniert den Passivhaus-Ansatz und eine neue Energieversorgungsmethode* mit dem Ziel, bei der Nachrüstung bestehender Gebäude eine 70-prozentige Reduzierung und bei Neubauten eine 90-prozentige Reduzierung bei vertretbaren Kosten zu erreichen. Die Referenzen sind die Baunormen für Wohnhäuser aus dem Jahr 2004 eines Landes.

Die beste Technologie allein kann den Klimawandel nicht beeinflussen. Die kanadische Vorstellung eines Passivhauses im Jahr 1978 scheiterte, weil die Kluft zwischen der fortgeschrittenen Bauwissenschaft und der Baupraxis zu groß war. Kanada und die USA reagierten darauf mit öffentlich-privaten hochwirksamen nationalen Pro-

https://doi.org/10.1515/9783112217023-203

grammen, die einen Paradigmenwechsel von der Verbesserung von Materialien bis zur Planung des gesamten Gebäudes und der Auswahl von Materialien für bestimmte Teile des Gesamtgebäudes bewirkten.

## 1.1 Spezialwissen zu Energie und Innenraumklima

Prof. Hutcheon (Hutcheon, 1998) definierte die Bauwissenschaft als einen Komplex, den man braucht, um die Vorhersagbarkeit einer Leistung zu erreichen, und er stellte fest, dass die leistungsbegrenzende Bedingung, der Misserfolg, ebenfalls bekannt sein muss. Der Abstand zum Misserfolg wird zu unserem Maß für die Leistung.

> *Das Wissen über das Bauen, das der Einfachheit halber als Baukunde bezeichnet wird, ist vor allem deshalb wertvoll, weil es nützlich ist, um das Ergebnis einer bestimmten Bausituation vorherzusagen ... Nur mit Wissen ist es möglich, die Relevanz von Erfahrungen zu beurteilen und somit auf eine breitere und vielfältigere Erfahrung zurückzugreifen, um eine Vorhersagbarkeit zu entwickeln.*

So ersetzte der Begriff „Gebäudewissenschaft" den Begriff „Gebäudetechnik", da das Ingenieurwesen Theorie und Praxis miteinander verbindet. In den USA praktizieren Ingenieure und Architekten, die als Berater für die Industrie tätig sind, die Gebäudewissenschaft. Sie wird nur selten an Universitäten gelehrt, da ihr multidisziplinärer Charakter auch einen breiten, multidisziplinären Hintergrund erfordert.

Bauphysik-Kurse beinhalten häufig eine vereinfachte Einführung in die Bereiche Umweltkontrolle, Akustik und Beleuchtung. Einem solchen Kurs fehlt der Bezug zur Baupraxis und zu den wirtschaftlichen Aspekten des Bauens, was zu einer Kluft zwischen der gelehrten und der in der Praxis erreichten Energieeffizienz führen kann.

## 1.2 Eine Kluft zwischen angestrebter und erreichter Energieeffizienz in der Praxis

Das American Institute of Architects hat eine kohlenstoffneutrale Zukunft ausgerufen, doch die Praxis zeigt, dass Einsparungen bei der Raumenergie für den Komfort der Bewohner genutzt werden und nicht zu einer Verringerung des Gesamtenergieverbrauchs führen. Wie aus Tabelle 1 hervorgeht, wurden im Zeitraum von 1978 bis 2005, als die Raumheizungsenergie reduziert wurde, die Komfortkomponenten erhöht (Bomberg and Onysko, 2008)). Die Bauherren nutzten die Einsparungen zum Zweck der Verbesserung von Komfort, und der Gesamtenergieverbrauch blieb gleich.

**Tabelle 1:** Der Gesamtenergieverbrauch (10,6 Billionen BTU) hat sich nicht verändert, die Raumheizung wurde reduziert, aber die Komfortkomponenten wurden erhöht (Bomberg and Onysko, 2008)).

| Kategorie | % im Jahr 1978 | % im Jahr 2005 | Unterschied |
|---|---|---|---|
| Raumheizung | 66 | 41 | −25% |
| Warmwasserbereitung | 14 | 20 | +6% |
| Klimatisierung | 3 | 8 | +5% |
| Haushaltsgeräte/Elektronik | 17 | 31 | +14% |
| Gesamtverbrauch an Energie | 100 | 99.7 | −0,3% |

## 1.3 Das Scheitern des ersten Passivhauses der Welt

Das Passivhaus aus dem Jahr 1978 war mit einer ähnlichen Technologie ausgestattet wie das heutige (Energy Conservation Houe, 1978). Es scheiterte jedoch, weil die Kluft zwischen fortgeschrittener Wissenschaft und Praxis zu groß war. Die Bauherren wünschten sich ein luftdichtes, gut isoliertes Haus, aber um die Kosten zu senken, ersetzten sie die Gasheizung durch eine elektrische Fußleistenheizung. Durch den Wegfall des Schornsteins änderten sie die Luftströmung und das Raumklima, was zu „ungesunden Gebäuden" (zu wenig frische Luft) und feuchten Dachböden (erhöhte Feuchtigkeit und Kondensation in den oberen Stockwerken und auf den Dachböden) führte. Dieser Misserfolg führte zu einer raschen Konsequenz für die kanadischen Bauvorschriften – zunächst wurde eine Empfehlung ausgesprochen, und 1985 wurde eine mechanische Belüftung in Kanada verbindlich vorgeschrieben. Die Probleme wurden im Rahmen eines Projekts im US-Bundesstaat New York (Bomberg et al., 2009, Brennan et al., 2008) angegangen, bei dem eine effiziente mechanische Belüftung als Hybridlösung mit beweglichen Hochgeschwindigkeitsrohren mit kleinem Durchmesser (Walburger et al., 2010) eingesetzt wurde.

Die Kluft zwischen Wissenschaft und Praxis veranlasste Nordamerika (USA und Kanada), nationale öffentlich-private Programme einzuführen und die Öffentlichkeit mit erheblichem Aufwand zu unterrichten. Diese Programme haben die Situation in Nordamerika verändert, und die Autoren sind der Meinung, dass es einer Neuauflage der Programme bedarf, die mit einer Vision für die Nachrüstung bestehender Gebäude kombiniert werden sollte (siehe weiter unten).

## 1.4 Das Scheitern des traditionellen Nachrüstungskonzepts

Abbildung 1 des Lawrence Berkeley National Laboratory (LBNL, 2008, (Bomberg and Onysko, 2008)) zeigt BAU, d. h. Bauen wie bisher (building as usual; ohne Änderungen) und schlägt für neue gewerbliche Gebäude eine Energiereduzierung auf 10 % und für alle Bestandsgebäude auf 50 % des Niveaus der frühen 2000er Jahre vor. Heute, 17

Jahre später, sind neue Gebäude auf dem aktuellen Stand, aber die Nachrüstung des Bestands ist ein überragender Misserfolg. Neue Gebäude werden als System konzipiert, während die Nachrüstung bestehender Gebäude *traditionell fragmentiert* ist und vor allem mit Blick auf die Rendite stattfindet.

Die Einteilung in Abbildung 1 ist rein verwalterisch. Der wirtschaftliche Wert des Gebäudes hängt von der Lage, den Funktionen und dem Wohnkomfort ab.

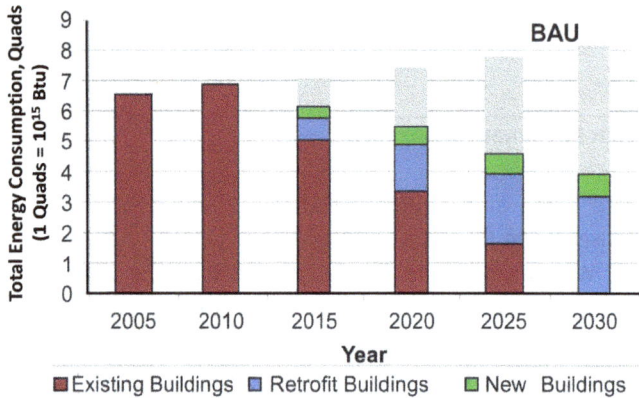

**Abbildung 1:** Um den Grenzwert für die Klimaerwärmung einzuhalten, prognostizierte das LBNL, dass neue Gebäude auf 10 % und nachgerüstete Gebäude auf 50 % der Energiecodes von 2004 begrenzt werden sollten. (Bomberg und Onysko, (Bomberg and Onysko, 2008)).

## 1.5 Die Notwendigkeit für eine Nachrüstungsvision

Wenn wir den Stand der Technik bei der Nachrüstung betrachten, dann sehen wir einige hervorragende Technologien, aber keine davon ist allgemeingültig. Passivhäuser in Deutschland unterscheiden sich von denen in Texas, aber die in Texas unterscheiden sich auch von denen in Illinois. Einige der Unterschiede haben mit dem Klima der Gegend zu tun, aber die meisten sind auf die jeweiligen Konventionen zurückzuführen. Bedenkt man die wachsende Kluft zwischen der Produktivität in der Industrie und der im Bauwesen, ist die Frage erlaubt, warum die industrielle Fertigung beim Hausbau nicht schneller voranschreitet. Außerdem hat die Inflation der Post-COVID-Zeit dazu geführt, dass die Kosten für Wohnraum für die junge Bevölkerung unerschwinglich geworden sind.

Mit Blick auf die Autobranche kann man feststellen, dass die technologische Revolution dem Paradigmenwechsel folgte. Schon Henry Ford stellte fest, dass das Ergebnis besser ist, wenn das Auto zum Arbeiter kommt. Das Konzept der Montage wurde zur Regel; die Massenproduktion senkte den Preis des Model T in 10 Jahren um den Faktor 5. Die zweite wissenschaftliche Revolution (Kuhn, (Kuhn, 1970)), die so genannte Qualitätsrevolution, beruhte auf der japanischen Beobachtung, dass wir, wenn wir nicht wissen, warum Menschen Fehler machen, ihnen helfen sollten, Fehler zu vermeiden, anstatt zu untersuchen, wie viele Fehler gemacht wurden, wie es die westliche Welt seit den 1950er Jahren getan hat.

Wir brauchen also eine Vision für die Nachrüstung. Dies war auch eine Schlussfolgerung der Konferenz für Bauphysik in Nanjing im Jahr 2010, denn das Klima in dieser Region der Welt ist das ganze Jahr über feucht. Die Kontrolle der Luftfeuchtigkeit allein ist einfach. Leider wurde die Bauwissenschaft in Amerika aber von der Politik beeinflusst (Bomberg et al., (Bomberg et al., 2020)), und in Europa wurde das Konzept der Luftumwälzung zugunsten der „grünen Vereinfachung" gestrichen. Die Lehren aus den Pandemien haben dazu geführt, dass die Luftfilterung in den Vordergrund der Überlegungen zum Innenraumklima gerückt ist. Wasser, dessen Wärmespeicherkapazität viermal höher ist als die der Luft, muss als Träger von Wärmeenergie erhalten bleiben, aber die Luftzirkulation, die bei den meisten Umweltüberlegungen eine Rolle spielt, ist die zweite Schlüsselkomponente der Energie- und Umweltplanung.

In der Tat muss eine Vision diese beiden Teilsysteme in den Prozess von Monitoring und Modellierung zur Bewertung und Verbesserung der Leistung neuer Technologien für nachgerüstete Wohnungen und Gebäude mit einbeziehen.

Es gibt noch einen weiteren Grund, aus dem die Wissenschaftsgemeinde eine Vision für die Renovierung, Sanierung, Aufrüstung, Modernisierung von Gebäuden und viele andere Begriffe, die von Politikern ohne Verstand verwendet werden, veröffentlichen muss, denn in den meisten Fällen bremsen grüne Subventionen den Fortschritt der Technologie auf dem Markt, anstatt ihn zu erleichtern. Ein Beispiel: Die letzte polnische Regierung unterstützte sowohl die Hersteller von Wärmepumpen als auch deren Kunden. Das Ergebnis war ein enormer Anstieg des Marktpreises für Wärmepumpen und die Verbreitung der schlechtesten technischen Lösungen, wie z. B. ein so genannter „wärmepumpenbetriebener Warmwasserspeicher", der zu 81 % aus einer direkten elektrischen Heizung und zu 19 % aus einer Wärmepumpe besteht. Wenn es eine wissenschaftlich belegte Vision für die Nachrüstung gäbe, könnte das Ausmaß der politischen Ignoranz verringert werden.

## 1.6 Verbesserung der Projektvision: Das Integrierte Designprotokoll (IDP)

In den 1990er Jahren wurden durch einen Kompromiss Unsicherheiten bei den Planungszielen verringert. Da die Vorhersage des Energieverbrauchs die Fähigkeiten von Architekten und Bauingenieuren überstieg, wurde das Planungsteam um einen Experten für Energiemodellierung erweitert. Gleichzeitig wurde die Analyse der Umweltbedingungen in Gebäuden in ein konzeptionelles Stadium verlagert. Dieser Prozess wurde als Planungs-Charette bezeichnet. Während der Französischen Revolution war die Charette der Wagen, der die Verurteilten zur Guillotine brachte. Das IDP-

Konzept verbreitete sich weltweit. Und warum? Liegt es daran, dass IDP für das Planungsteam eine gemeinsame Vision des Gebäudes schafft? Vielleicht, aber der wahre Grund war wirtschaftlicher Natur. Indem die Umweltentscheidungen im Vorfeld getroffen wurden, reduzierte IDP die Kosten für die Planung.

## 2 Universelle Gebäudeenergie- und Umwelttechnik

Die älteste Forschungsschrift, ein vertraulicher Bericht des National Research Council of Canada (Bericht 1639 vom Dezember 1947), enthielt Feldmessungen im Winter 1946/7 an zwei einstöckigen Gebäuden, in denen Metallrohre unter dem gesamten Erdgeschoss auf einem teilweise isolierten Betonfundament verlegt waren, um das Gebäude mit einer Warmwasserheizung zu versorgen. Jedes Gebäude hatte einen 1,2 m hohen Kriechkeller über der beheizten Decke und einen Wohnraum von etwa 20 m². In dem Bericht wird die dem Gebäude zugeführte Wärmemenge in Abhängigkeit von der Außentemperatur und dem Belüftungsniveau dargestellt (die Tests wurden in einem Belüftungsbereich zwischen 1,5 und 4 ACH durchgeführt).

Etwa 60 Jahre später verwendet ein ungarischer Erfinder einen Warmwasser-Wärmetauscher in der zusätzlichen Betonplatte, die sich etwas unterhalb des Gebäudes befindet. Der geothermische Warmwasser-Wärmetauscher in Ungarn lieferte nicht genug Energie, so dass das Belüftungssystem um eine Wärmepumpe unter dem Kellergeschoss ergänzt wurde. Man nannte es „aktive Wärmedämmung" (Kisilewicz et al., (Kisilewicz et al., 2019, Kisilewicz et al., 2023)). Zur gleichen Zeit baute ein Doktorand an der Universität Syracuse (. Lingo and Roy, 2016) einen Wärmetauscher, der die Luft in den Wänden über eine Split-Level-Wärmepumpe zirkulieren lässt, die mit einem geothermischen Wassertank im Boden betrieben wird, und wechselte zu wasserbasierten Wärmetauschern.

Etwa zur gleichen Zeit erkannte ein Team von motivierten Menschen in der kanadischen Stadt Montreal, dass nur die Berücksichtigung vieler Designdetails zum Erfolg führt, und formulierte den Integrationsprozess als mehrstufigen Bau (12). Im Laufe von 10 Jahren schrittweiser Konstruktion reduzierten sie den Energieverbrauch im Jahr 2018 um 90 Prozent, wie vom Lawrence Berkeley National Laboratory (LBNL, siehe Konferenz BEST-1, 2008 (Bomberg and Onysko, 2008)) ausgerufen.

Die Notwendigkeit einer allgemeingültigen Technologie, die auf einer westlich-chinesischen Bauwissenschaft-Konferenz im Jahr 2010 in Nanjing deutlich gemacht wurde, führte zu einem kleinen virtuellen Netzwerk (Mattock, 2010, Bomberg and Anna, 2021, Thorsell and Bomberg, 2011, Bomberg et al., 2015, Bomberg et al., 2016a), das schon an der thermischen Massewirkung in Holzständerhäusern gearbeitet hatte (Mattock, 2010)), ein Musterhaus des ungarischen Erfinders wurde 2008 gebaut (Kisilewicz et al., 2019, Kisilewicz et al., 2023), ein Musterhaus des amerikanischen Doktoranden im Jahr 2008 (Lingo and Roy, 2016), in Montreal baute man von 2008-2018 mehrstufig (Rosemount Atelier in Montreal, 2016), die Technische Universität

Krakau arbeitete an neuronalen Netzen (Dudek et al., 2020, Dudzik et al., 2020, Dudzik, ) und der Vortemperierung der Lüftungsluft (Romańska-Zapała et al., 2018, Fedorczak-Cisak et al., 2022, Bomberg et al., 2016) in Zusammenarbeit mit dem US-Passivhaus-Institut (Klingenberg et al., 2009, Wright and Klingenberg, 2015), der Tech. University of Quebec (Heibati et al., 2019, Heibati et al., 2021) und der Southeast University (Hu et al., ). Im japanischen Tokio werden mit einem Verfahren, das 2020 den amerikanischen Technologiepreis gewann und das als thermoaktives System bezeichnet wird (Hu et al., ), sowohl die Warmwasserheizung/-kühlung als auch die Lüftungsluft für die Energiezufuhr (oder -abfuhr) verwendet. Die 1999 begonnene und weitergehende Entwicklung im Bereich der Feuchtigkeitsregulierung (Häupl et al., 1999) und des Innenraumklimas (Fort et al., 2020, De Deer and Zhang, 2018) vervollständigen die erste Phase unserer Netzwerkaktivitäten.

Das Projekt Atelier Rosemount im kanadischen Montreal (Rosemount Atelier in Montreal, 2016) hat durch die Einführung der mehrstufigen Bauweise die Hürde zur Bezahlbarkeit genommen. In dem Bauviertel, in dem Luxuswohnungen Sonnen- und Schattenseiten haben, um Querlüften zu ermöglichen, befinden sich auch Sozialwohnungen, deren Kosten nur einen Bruchteil ausmachen, die aber ungefähr den gleichen Komfort bieten. Das Finanzierungskonzept, das mit diesem Projekt eingeführt wurde, bestand aus einigen kurzfristigen Darlehen für jede neue Bauphase. In unserem Buch wird das „zweistufiger Bauprozess" genannt.

Der ungarische Erfinder, dessen Messdaten von Kisilewicz et al. analysiert wurden (Kisilewicz et al., 2019, Kisilewicz et al., 2023), führte in Zusammenarbeit mit der Universität Syracuse (NY, USA) den aktiven Teil des passiven und thermoaktiven Clusters (PTAC) ein, zu dem das amerikanische Lawrence Berkeley National Laboratory (LBNL) (Bomberg and Onysko, 2008) eine Quantifizierung für die Vision von nachhaltigem Bauen lieferte. Die Synthese der Arbeiten von Atelier Rosemont, Montreal, Kanada, des ungarischen Erfinders, des Tokioter Designteams (Kosuke et al., 2020) und des LBNL lässt sich wie folgt zusammenfassen:

*Die Grundlage einer universellen Technologie für nachhaltige Gebäude mit nahezu Null-Energie und nahezu Null-Kohlenstoff-Emissionen wurde im Jahr 2020 geschaffen.*

Trotz dieser und anderer Veröffentlichungen (Buratti et al., 2015, Ferreira et al., 2012) sind die mehrstufigen oder thermoaktiven (TA) Technologien bisher nicht weitverbreitet. *Dies impliziert eine schlechte Verbindung zwischen der Wissenschaftsgemeinde, die im Bereich Energie und Innenraumklima arbeitet, und der Baupraxis.*

## 3 Universelle Energie- und Umwelttechnik für das Nachrüsten

Investoren müssen sich bei Vorschriften und Normen an die Mindestanforderungen halten. Die Gesellschaft braucht jedoch ein viel höheres Investitionsniveau, entweder Netto-Nullenergie oder zumindest Fast-Nullenergie. Der zweistufige Bauprozess entschärft diesen Konflikt. Die zweite Phase des Projekts unterliegt denselben finanziellen Beschränkungen wie jedes andere Sanierungsprojekt. Sowohl Bauherren als auch Hauseigentümer, die eine Hypothek beantragen, sehen sich mit zwei kritischen Fragen konfrontiert: (a) dem Wert der bestehenden Immobilie im Vergleich zu den umliegenden Grundstücken und (b) den geschätzten Kosten für die geplanten Reparaturen. Ein Kostenvoranschlag aus Phase 1 ist von unschätzbarem Wert. Der nächste Abschnitt berichtet über ein erfolgreiches, mehrstufiges Bauprojekt.

### 3.1 Mehrstufiges Bauen: Atelier Rosemont in Montreal, Kanada

Der erste Bauabschnitt begann im Jahr 2008, und es wurde schrittweise weitergebaut, bis 10 Jahre später 92 % der kumulativen Reduktion erreicht waren. Die Nachrüstung umfasste die folgenden Schritte:
- Hochleistungsgehäuse; gemeinschaftliches Wasser; Solartechnik, die zu einer Reduzierung um 36 % führt.
- Abwasser, die passiven Maßnahmen zur Energieeinsparung ergeben 42 %.
- Wärmepumpenheizung – alle passiven Maßnahmen führen zu einer Reduzierung um 60 %.
- Warmwasserbereitung mit Vakuumröhrenkollektoren, weitere 14 %.
- Photovoltaik-Paneele reduzieren die Gesamtenergie auf insgesamt 92 %.

Ein mehrstufiger Bauprozess, wie er im „Atelier Rosemont" in Montreal (Rosemount Atelier in Montreal, 2016) eingesetzt wurde, beseitigte den Unterschied zwischen Neubau und Nachrüstung, indem die Nachrüstung auf die nächste Stufe jedes Neubaus übertragen wurde. Die oben dargestellten Energieeinsparungen stimmen mit den Erfahrungen aus den verschiedenen Anwendungen des Building America-Programms überein (Private Gespräche mit IBACOS- und BCI-Teams der BA, 2018).

### 3.2 Integration ermöglicht neue Methoden der Energieversorgung

Aufgrund des Energievervielfachungseffekts sind Wärmepumpen die bevorzugte Wahl. Es gibt allerdings verschiedene Arten von Wärmepumpen, und eine effiziente Nutzung der neuen Technologie erfordert den Einsatz einer Wasserwärmepumpe zusammen mit zwei Wassertanks als Wärmespeicher (Wärmekapazität). In den Wassertank wird eine Heizspirale eingesetzt, die heißes Wasser an die Boden- und Wandwärmetauscher liefert. Die Wassertanks werden wie folgt bezeichnet: (1) Warmwasserspeicher und (2) Kaltwasserspeicher, wobei letzterer als unterer Anschluss der Wärmepumpe fungiert.

  Es wird zwischen zwei Arten von Wärmespeichern unterschieden: Kurzzeitspeicher mit einer thermischen Kapazität von 14 bis 18 Stunden zum Ausgleich der täglichen Lasten (Fadiejev et al., 2017) und Langzeitspeicher mit einem wöchentlichen

Ausgleich der extremen thermischen Lasten (Bomberg, 2021). Wasserbasierte Wärmepumpen (WSHP) haben in der Regel einen höheren Leistungskoeffizienten als luftbasierte Wärmepumpen. Bei der Umstellung auf ein wasserbetriebenes System muss man jedoch die Temperaturuntergrenze bedenken. Wenn beispielsweise die Temperatur im Kaltwassertank unter 10 °C sinkt, beginnt eine elektrische Heizung, dem Wasser Energie zuzuführen. Die elektrische Heizung hat jedoch einen angenommenen Leistungskoeffizienten von eins, während wasserbasierte Wärmepumpen einen Leistungskoeffizienten von mehr als 4 haben können. Es ist sinnvoll, warmes Wasser anstelle von elektrischer Heizung zu verwenden, um dem Kaltwassertank Wärmeenergie zuzuführen. Wir verwenden einen 40-Liter-Zusatzwassertank und einen 120-Liter-Abwassertank (Mindestbedarf für 2 Erwachsene pro Tag für Sanitär- und Warmwasserverbrauch) und die folgende Routine.

Wenn die Temperatur im Kaltwassertank unter 10 °C fällt oder wenn jemand zu duschen beginnt, werden 40 Liter aus dem Kaltwassertank in den Zusatzwassertank und später in den Abwassertank geleitet. Der Abwassertank dient der Wiederverwertung von Wasser aus dem Kaltwassertank, wenn die Temperatur unter oder über den Richtwert für den Kühlbetrieb fällt. Derselbe Vorgang wird in kalten Klimazonen durchgeführt, wenn dem Kaltwassertank warmes Wasser aus einem Bad oder einer Dusche zugeführt wird. Auf diese Weise gelangt im Winter kaltes, recyceltes Wasser in den Abwassertank, und im Sommer, wenn die wasserbasierte Wärmepumpe zur Kühlung der Wohnung eingesetzt wird, wird heißes Wasser zum Spülen der Toilette verwendet. Es mag paradox klingen, aber da das Hinzufügen eines Abwassertanks das Volumen der Wasserentnahme senkt und gleichzeitig zur Optimierung des Leistungskoeffizienten beiträgt, macht es den Betrieb des Systems weniger teuer.

### 3.3 Die Wandwärmetauscher: Paneele oder Einbau vor Ort

Bei der großen Fläche von Wandwärmetauschern wird mit 55 °C eine niedrige Temperatur im Heizungssystem für Wohnräume gewählt. Das Wandsystem ist etwa 25–30 % wirtschaftlicher als ein Fußbodensystem. Außerdem wird durch die Verwendung einer anpassungsfähigen Innentemperatur der Masseneffekt verstärkt (Fadejev et al. (Bomberg, 2021)). Diese Werte wurden mit der Energy Plus-Computersoftware mit typischen Folienkoeffizienten für horizontale und vertikale Ausrichtungen berechnet. Darüber hinaus stellte Hu (Covey, 1989) fest, dass man die PEX-Rohre auf einer Wärmedämmung mit einem Wärmewiderstand von mindestens 1 $(m^2 \cdot K)/W$ platzieren sollte, wenn man eine 90 %ige Heizleistung erreichen will..

Es gibt zwei Methoden zum Bau von Wärmetauschern. Werden sie vor Ort verbaut, sind die PEX-Rohre durchgehend. Bei einer Paneellösung werden die PEX-Rohre mit Schnappverlüssen verbunden. Paneele können aus verschiedenen Materialien hergestellt werden. In China beispielsweise wurden Paneele mit MgO-Zement und gefrästem (faserigem) Reis- und Holzabfall hergestellt, um sowohl die Feuchtigkeit zu puffern als auch die Elastizität zu gewährleisten.

**Wahl des Wärmepumpentyps für die Heizungsanlage.** Die Wahl der Wärmepumpe richtet sich nach der lokalen Konvention. In Amerika ersetzen Luftwärmepumpen die alten Fensterklimageräte, und auch in Europa nutzen fortschrittliche Konvektions-Strahlungs-Wärmetauscher die außen aufgestellte Wärmepumpe. Der Nachteil dieser Lösung ist jedoch, dass der Ventilator ständig in Betrieb ist und der Leistungskoeffizient in der Regel unter 3 liegt.

Umgekehrt bietet eine WSHP mit einem Leistungskoeffizienten von über 4,0 einige Vorteile: gleichzeitiger Zugang zu heißem und kaltem Wasser das ganze Jahr über, einfache Integration von Wasserwiederaufbereitung, Einbeziehung von hybriden Sonnenkollektoren und Belüftungssystemen, Möglichkeit geometrischer Erweiterung des Energiesystems, mehrere Heiz- und Kühlstufen und schließlich Integration in ein neues „Fernklimanetz". Der Nachteil ist der typischerweise laute Kompressor der geothermischen Wärmepumpe. Dieses Problem muss noch gelöst werden.

### 3.4 Andere Verbesserungen, die in der PTAC-Technologie enthalten sind

Steven Covey betonte in „7 Habits of the Most Efficient People", dass jeder Prozess mit der Definition des gewünschten Ergebnisses beginnen sollte (Covey, 1989). Um ein Gleichgewicht zwischen den oft widersprüchlichen Anforderungen der Energieeffizienz, des Nutzerkomforts und der hohen Qualität des Innenraumklimas zu erreichen, muss im Integrierten Designprotokoll (IDP (IDP, 1991)) sowohl analoges als auch analytisches Denken eingesetzt werden. Das IDP schuf eine gemeinsame Vision für die Planung auf der Grundlage eines ganzheitlichen Gebäudeansatzes, reduzierte die Kosten für die Planung selbst und verlagerte das Planungsparadigma auf das Gebäude als System. Die Planung des gesamten Gebäudes mit primären Anforderungen an Komponenten und Baugruppen beinhaltete die Wahl der Materialien in einer späteren Phase, was die Polymerindustrie in die Lage versetzte, neue multifunktionale (Bomberg et al., 2017) Materialien zu entwickeln.

Im kalten städtischen Klima Montreals reduzierte ein hohes Maß an Wärmedämmung und Luftdichtheit den Energieverbrauch um etwa 40 %. Eine Luftwärmepumpe wurde zu den passiven Baumaßnahmen im Staat New York hinzugefügt, und die Energie wurde um 55 % reduziert (siehe Bauprozess (Walburger et al., 2010) und Qualitätssicherung (Brennan et al., 2008)). Ein Doktorand, der die Co-Simulation als Verbesserung der Energiemodellierung einsetzte, zeigte, dass im kalten Klima die passiven Maßnahmen durch den Einsatz einer Wärmepumpe auf 60 % Reduktion begrenzt sind (Heibati et al., 2021). Um mehr als 60 % zu erreichen, müssen zusätzliche Maßnahmen ergriffen werden, z. B. Geosolartechnik oder eine Änderung des Energieversorgungssystems. Das im Jahr 2010 begonnene Konzept (Bomberg, 2010) mit den dokumentierten Arbeitsfortschritten (Romanska-Zapala et al., 2018, Yarbrough et al., 2018, Romanska-Zapala et al., 2018) ist nun in der PTAC-Systemtechnik enthalten.

Der Prozess der Luftfilterung, der die Ausbreitung der mikrobiologischen Verschmutzung kontrolliert, erfordert eine Belüftung mit variabler Rate. Dies kann mit Luftspalten zwischen bestehenden Wänden und einem neuen Heiz-/Kühlsystem durch Erweiterung des Nachrüstungssystems erreicht werden. PTAC (passive thermoaktive Cluster) ist eine universelle, klimaangepasste, öffentlich nutzbare Technologie, die zu einem logistischen Dach entwickelt wurde, unter dem verschiedene Optionen wie künstliche neuronale Netze mit kommerziellen Service- und Komfortsystemen interagieren können. Wir schlagen keine spezifische, detaillierte Technologie vor, sondern eine Blaupause für verschiedene praktische Optionen. Es ist Sache des Planers, die Kapazität des Wassertanks, die Leistung der wasserbasierten Wärmepumpe (WSHP), die nur in der Nacht arbeitet, oder die Verlängerung der Betriebsstunden der WSHP, die Verwendung eines ein- oder zweistufigen Wasserpuffersystems und Maßnahmen zur Sicherstellung der Mindesttemperatur des Niederdruckterminals auszuwählen. Wenn das Versorgungssystem zu teuer ist, kann der Planer die Wärmedämmung der zusätzlichen Wände erhöhen oder Phasenwechsel-/Reflexionsflächen in den Heiz-/Kühlpaneelen einsetzen. Man kann auch die Nachrüstungsfläche vergrößern und alle inneren Trennwände abdecken.

Das nächste Element zur Erweitung des Arbeitsbereichs ist nicht offensichtlich. Aus Platz- und Preisgründen kann man sich nicht auf die Geothermie verlassen. Stattdessen kann man sich auf den Leistungskoeffizienten von WSHP verlassen, insbesondere bei der Regelung der Temperatur des unteren Anschlusses der WSHP. Außerdem kann etwa 1/3 der Wärmeenergie des Duschwassers leicht zurückgewonnen werden, wenn wir das Abwasser in das System miteinbeziehen. Die nächste Änderung in der PTAC-Technologie ist die Umstellung von einer Luft-Wasser-Wärmepumpe auf eine Wasser-Wasser-Wärmepumpe, wobei zwei Wassertanks und eine unabhängige Wasserpumpe verwendet werden, um die Wasserheizung/-kühlung von der Raumklimatisierung zu trennen. Der Hauptgrund dafür ist die Erhöhung des Leistungskoeffizienten von 2,5 bis 3 für die herkömmliche Wärmepumpe auf über 4 für die wasserbasierte Wärmepumpe (WSHP). Weitere Vorteile der WSHP sind die ganzjährige Verfügbarkeit von Warm- und Kaltwasser, die einfache Integration von hybriden Sonnenkollektoren, Abwasser und Fernklimanetzen sowie die Erhöhung der Heiz- und Kühlleistung.

Das Rückgrat der PTAC-Technologie ist die Modellierung und Leistungsbewertung (MAPE), bei der die Feldmodellierung die Kalibrierung für numerische oder neuronale Netzmodelle liefert, die wiederum für den Aufbau automatisierter Kontrollsysteme verwendet werden können. Dennoch wird die Technologie allein ohne die spezifischen sozialen Anforderungen das Problem nicht lösen. Wie wir oben gezeigt haben, wurden die besten Technologien vom Markt nicht angenommen. Der Bauherr wird das tun, wofür er bezahlt wird. Wir wissen, dass eine wissenschaftliche Revolution notwendig ist, damit die Gesellschaft die Kontrolle über die Geschwindigkeit des Klimawandels erlangt, die Wirtschaft lokale Arbeitsplätze schafft und die Bewohner die Qualität der Innenraumumgebung verbessern können.

### 3.5 Senkung der Kosten der Dekarbonisierung

Torrie und Bak (Torrie and Bak, 2022) hoben für Kanada hervor, dass das Land trotz zehn Millionen Einfamilienhäusern und fünf Millionen Wohnungen mit einer Fläche von 2,1 Milliarden Quadratmetern und 65 Millionen Tonnen Kohlenstoffemissionen pro Jahr, von denen etwa zwei Drittel mit Erdgas betrieben werden, nicht auf dem Weg in eine kohlenstoffarme Zukunft ist. Früher wurde jedes Fahrzeug einzeln hergestellt, so wie wir es heute mit Wohnungen machen. In heutigen Dollar ausgedrückt, kostete ein Ford 25.000 US-Dollar, und das Unternehmen verkaufte davon 19.000 Autos. Zehn Jahre später sanken die Kosten für das Model T auf 5.000 US-Dollar, und Ford verkaufte 941.000 Fahrzeuge.

Im Jahr 2019 gaben die Kanadier jährlich mehr als 60 Milliarden Dollar für die Renovierung ihrer Häuser und 30 Milliarden Dollar für die Raumheizung aus. Eine typische tiefgreifende Nachrüstung mit Umstellung auf eine Wärmepumpe für ein Einfamilienhaus kostet mindestens 40.000 $. Berechnet man einen 10-jährigen Übergang zu einem Niedrigenergieniveau mit durchschnittlich 36,7 Mrd. $ pro Jahr, so ergeben sich Kohlenstoffemissionskosten von 141 $ pro Tonne (etwa 100,00 US$). Wenn man davon ausgeht, dass eine neue Technologie den Preis um 30 % senkt, wenn die Kohlenstoffeffizienz um 30 % steigt und das Volumen der Nachrüstung um das Dreifache wächst, sinken die Kosten für eine Tonne Emissionen auf unter 10 $ pro Tonne.

In jüngster Zeit erlebte die Automobilindustrie eine weitere Revolution (eine Qualitätsrevolution). Das in Japan entwickelte Qualitätssicherungssystem mit kontinuierlicher Qualitätsverbesserung ist das verbindende Ziel der Fertigungsorganisation. Da wir uns heute mitten in der 4. industriellen Revolution befinden, sollten wir die Umrüstungstechnologie neu überdenken.

Die obige Diskussion definiert unsere Ziele. Wir müssen die Energieeffizienz bei gleichbleibenden Kosten um 1/3 steigern und das Volumen der Nachrüstung erhöhen, um die Kosten für 1 Tonne CO2 um den Faktor 10 zu senken. Dies ist ein Ziel, um die Geschwindigkeit des Klimawandels durch die Nachrüstung bestehender Gebäude zu verlangsamen.

### 3.6 Erfindung eines Fernklimasystems als Ersatz für die Fernwärme

Diese EQM-Technologie ist eine Innovation im Fernklimanetz, die auch bei historischen Gebäuden angewendet werden kann, indem sie mit einem benachbarten Standardgebäude gekoppelt werden. Auf diese Weise werden die beiden Gebäude in ein lokales Fernwärme-/Kühlsystem einbezogen. Fernwärme-, Fernkälte- und Fernlüftungssysteme heben die Unterschiede zwischen einzelnen Gebäuden oder Stadtteilen auf. Da in der EQM-Technologie thermische Speicher und Wassertanks unterirdisch platziert werden können, kann das Fernklimasystem entweder Teil des Gebäudes oder des Energieverteilungssystems sein.

Die Forschung zu Luft-Erdwärmetauscher (Romanska-Zapala et al., 2018) ist zum Ergebnis gekommen, dass eine Tiefe von 1 m in Mitteleuropa günstig ist. Wenn ein Polyurethanschaum niedriger Dichte (ca. 10 kg/m$^3$) diese Leitung ausfüllt, ist der Schaum im Winter eine trockene und im Sommer eine feuchte Isolierung, im Sommer ein Wärmeleiter (Gefälle nach innen) und leitet daher die Wärme besser ab als ein trockener.

Zusammenfassend lässt sich sagen, dass die Weiterleitung des Rücklaufwassers zusammen mit der vorklimatisierten Luft in das nächste Gebäude die Betriebskosten senken kann, indem der Leistungskoeffizient gegenüber der der Wärmepumpe allein erhöht wird. Da der Bau von Siedlungen weniger kostspielig ist als der von Einzelgebäuden, erhöht die Hinzufügung eines Fernklimasystems die Effizienz der Investitionen und würde den Unterschied zwischen Einzelgebäuden und Stadtvierteln verkleinern.

## 4 Die Vorteile des vorgeschlagenen Ansatzes

Die Vorteile eines ganzheitlichen Ansatzes sind:
– Es gibt zwei Methoden des Systemaufbaus:
   (1) vor Ort, mit kontinuierlichem Aufbau von Hydronikkreisläufen, und
   (2) ein Paneelsystem mit Heizung und Kühlung, mit oder ohne Lüftung.
– Ein typisches Steuerungssystem benötigt eine Echtzeit-Energiemodellierung, während unsere Energiemodelle parametrisch sind, d. h. sie ermöglichen den Vergleich der Bedeutung verschiedener Parameter. Um sie für Echtzeitberechnungen zu verwenden, muss man das Modell an überwachten Felddaten kalibrieren, eine Co-Simulation von individuell verifizierten numerischen Modellen verwenden oder (c) kombinierte CFD- und numerische Modelle verwenden.
– Monitoring und Modellierung für die Bewertung (MAPE) ermöglichen es, Modelle zu kalibrieren und sie anschließend zur Optimierung mechanischer Geräte zu verwenden. Besonders erfolgreich sind einfache und präzise künstliche neuronale Netzmodelle.
– In den meisten Mittel- und Hochhäusern wird die Luftdruckkorrektur in den Fluren als Kamineffekt genutzt, der in jedem Stockwerk anders ist und je nach Jahreszeit variiert. Diese Unterschiede im Luftdruck haben einen erheblichen Einfluss auf die Luftqualität in Innenräumen. Die Hinzufügung von Luftdruckunterschieden macht einen kleinen Unterschied bei den Überwachungskosten, aber einen großen Unterschied bei der Analysefähigkeit, insbesondere wenn eine variable Luftwechselrate in die Gebäudeplanung einbezogen wird.

- Die Modellierung sollte sich auf eine lokale Wettervorhersage stützen, die anschließend mit den aus dem Monitoring erhaltenen Informationen abgeglichen wird. Die Modellierung des Innenraumklimas und der Energie in einem Hochhaus wird dadurch ziemlich kompliziert.
- Mit diesem Projekt wird auch eine neue Art von Fernklimanetz eingeführt (Bomberg, 2021). Das Rücklaufwasser von Gebäude eins wird in Gebäude zwei für den unteren Anschluss der Wärmepumpe verwendet. Dieses System kann für zwei Gebäude (historische Gebäude) oder die gesamte Gebäuderegion verwendet werden.
- Das Fernklimanetz befasst sich mit der Vorklimatisierung der Luft und verringert den Unterschied zwischen der Planung eines einzelnen Gebäudes oder der gesamten Siedlung und eröffnet damit neue Möglichkeiten der Abwägung.
- Schließlich erfordert die Komplexität der Wechselwirkungen zwischen den Teilsystemen in Mittel- und Hochhäusern eine experimentelle Überprüfung und Analyse aller Komponenten des Systems in allen vier Jahreszeiten.

## Schlussfolgerungen

In diesem Buch erweitern wir die Passivhaus-Methodik durch die Verwendung folgender Aspekte:
- Zweistufiger (mehrstufiger) Bauprozess, der die Finanzierungsmuster verändert.
- Gebäudeautomation zur Steuerung des Beitrags der thermischen Masse und zusätzlicher thermischer Wasserspeicher in Verbindung mit wasserbasierten Wärmepumpen und bei Neubauten auch mit Sonnenkollektoren.
- Anpassungsfähiges Innenraumklima durch eine in die Gebäudestruktur integrierte HLK-Anlage und das MAPE-System (Monitoring and Performance Evaluation), um Energie und Innenraumklima während des Betriebs des Gebäudes zu optimieren.
- Fernklimanetzwerke zur Verbindung des passiven und thermoaktiven Clusters (PTAC) mit dem nächsten Gebäude in einer Reihe von 2 bis 200 Gebäuden zur Aufwertung eines historischen oder städtischen Viertels.

Das Wort Cluster in der obigen Definition bedeutet sowohl eine Ansammlung von passiven und thermoaktiven Methoden als auch eine Ansammlung von Gebäuden oder im Extremfall sogar ein Gebäude mit umliegendem Gelände, in dem große Wassertanks gelagert werden könnten, oder auch nicht gelagert werden dürfen, wenn die Stadt bereits gebaut ist.

Für die Nachrüstung bestehender Gebäude ist eine technische Revolution wie das Modell T von Ford erforderlich. Diese Revolution wird eine Win-Win-Lösung für die Gesellschaft, die Wirtschaft und die Bewohner des Gebäudes darstellen. Die Gesellschaft gewinnt durch die Verlangsamung des Klimawandels, die Wirtschaft durch viele lokale Arbeitsplätze und die Bewohner durch ein erschwingliches, hervorragendes Raumklima. Da die Bauherren nur das tun, was die Gesellschaft von ihnen verlangt, sollte die Gesellschaft verlangen, dass die Gebäude keine Kohlendioxidemissionen aufweisen und die Bewohner einen höheren Wohnkomfort haben.

Produktivität und Wohlbefinden der Bewohner sind die Hauptkriterien, Energie hingegen ist nicht sichtbar. Wenn wir einen Aspekt von vielen betonen, erwecken wir den Eindruck einer unausgewogenen Technologie. Medows (Meadows, 1972) sagte, dass Zuschüsse, Steuererleichterungen und Sponsoring keine Auswirkungen auf eine nachhaltig gebaute Umwelt haben, während der höchste gesellschaftliche Wert wie der Klimawandel große Auswirkungen hat. Warum nutzen wir sie nicht?

Die Antwort ist einfach: Der globale Markt gehört niemandem. Die Werte der Gesellschaft werden im Streben nach globalem Geldgewinn zerstört, und wir müssen heute andere Wege gehen, um lokale sozioökonomische Werte wieder aufzubauen. In der Arbeitsgruppe für Bauphysik einer führenden Universität in Montreal wusste tatsächlich niemand von dem Rosemont-Projekt. Ohne die Beteiligung der Wissenschaftler wusste die Gesellschaft nichts von den Fortschritten, niemand hat sich für die Vorteile dieses Projekts eingesetzt. Man kann über die mangelnde Qualifikation dieser Leute sprechen, die ökologisches Bauen ohne Einbeziehung der Marktbedürfnisse einführen wollen. Was wir brauchen, sind keine Geldgeschenke, sondern eine ernsthafte Aufklärungsarbeit in Form von Workshops, Präsentationen und Berufsverbänden für Bauherren. Vor Jahren hatten wir all diese Instrumente, aber aufgrund des fehlenden systematischen Technologietransfers und der Tatsache, dass Politiker akademische Projekte vorantreiben, ohne sie auf dem Markt umzusetzen, haben wir den Kontakt zwischen der akademischen und der Bauwelt verloren.

Die Epidemien haben die alte Ordnung zerstört, und während die Sponsoring-Produkte auf dem Markt keinen Fortschritt bringen, muss die Sponsoring-Organisation anfangen, marktähnlich zu denken, um eine Marktnachfrage zu schaffen. Wir wiederholen: *Die Technologie ist nur ein Bein, zum Laufen braucht der Mensch zwei.* Die unmittelbare Lösung sollte darin bestehen, das Sponsoring-Geld für den Bau des fehlenden Teils zu verwenden, damit die Gesellschaft die Technologie nutzen kann, bevor sie veraltet und überflüssig wird.

## Referenzen

Bomberg, M.; Onysko, D. Energy Efficiency and Durability of Buildings at the Crossroads, 2008,
    http://thebestconference.org/BEST1 (Zugriff am 25. Februar 2020).
Hutcheon, N.B.; Der Nutzen der Bauwissenschaft, (Vortrag von 1971). *J. Build. Phys.* 1998, *22*, 4.
Energy Conservation Houe, 1978, Regina, Saskatchewan, Provincial Gov, Pamphlet
Bomberg, M.; Brennan, T.; Henderson, H.; Stack, K.; Walburger, A.; Zhang, J.High Environmental
    Performance (HEP), residential housing and building technology, for NY state. *Ein Abschlussbericht für
    die NY State Energy Research and Development Agency und das National Center of Energy Mgmt,*
    Manuskript, 2009, 2009, USA

Brennan, T.; Henderson, H.; Stack, K.; Bomberg, M. Quality Assur. and Commissioning Process in High Environmental Performance (HEP) Demonstration House in NY State. 2008. online: www.thebestcon ference.org/best1 (abgerufen am 12. Oktober 2019).

Walburger, A.; Brennan, T.; Bomberg, M.; Henderson, H. Energy Prediction and Monitoring in a High-Performance Syracuse House. 2010. Online verfügbar: http://thebestconference.org/BEST2 (abgerufen am 2. Oktober 2019).

Kuhn, T.S. The Structure of Scientific Revolution; The Chicago U. Press: IL, USA, 1970, siehe auch The Fourth Industrial Revolution | Essay von Klaus Schwab | Britannica, abgerufen am 17.12.2023

Bomberg, M.; Romanska-Zapala, A.; Yarbrough, D.W. History of American Building Science: steps leading to scientific revolution. *J. Energies*. 2020, 13(5), 1027.

Kisilewicz, T.; Fedorczak-Cisak, M.; Barkanyi, T. Aktive Wärmedämmung als Element zur Begrenzung von Wärmeverlusten durch Außenwände. *Energy Build.* 2019, *205*.

Kisilewicz, T.; Fedorczak-Cisak, M.; Sadowska, B.; Ickiewicz, I.; Barkanyi, T.; Bomberg, M.; Gobcewicz, E.. On the results of long-term winter testing of active thermal insulation. *Energy Build.* Oct 2023, 296, 113412.

Lingo, L. Jr; Roy, U. Novel Use of Geo solar Exergy and Storage Technology in Existing Housing Applications: Conceptual Study. *J. Energy Eng.* 2016, *143*, 0401602.

Rosemount Atelier in Montreal. *Informationsnotizen;* Kanadische Hypotheken- und Wohnungsbaugesellschaft: Ottawa, Ontario, Kanada, 2016.

Mattock, C. Projekt Harmony House Equilibrium. In Proceedings of the Canada Green Building Council, Annual Conference Vancouver, Vancouver, BC, Canada, 8–10 June 2010.

Bomberg, M.; Romanska-Zapala, A; Yarbrough, D. Auf dem Weg zu einem neuen Paradigma für die Gebäudewissenschaft (bldg physics). *World* . 2021, *2*(2), 194–215. doi: https://doi.org/10.3390/world2020013.

Thorsell, T.; Bomberg, M. Integrated methodology for evaluation of energy performance of the building enclosures. P3: Uncertainty in thermal measurements. *J. Build. Phys.* 2011, *35*, 83–96.

Bomberg, M.; Gibson, M.; Zhang, J.. A concept of integrated environmental approach for building upgrades and new construction: Pt 1-setting the stage. *J. Build. Phy.* 2015 2015, *38*(4), 360–385.

Bomberg, M.; Wojcik, R.; Piotrowski, Z.. A concept of integrated environmental approach, part 2: Integrated approach to rehabilitation. *J. Build. Phys.* 2016a, *39*, 482–502.

Dudek, P.; Górny, M.; Czarniecka, L.; Romanska-Zapała, A.. IT system for supporting the decision-making process in integrated control systems for energy efficient buildings. In Proceedings of the 5th Anniversary of World Multidisciplinary Civil Engineering-Architecture-Urban Planning Symposium-WMCAUS 2020, Prague, Czech Republic, 31 August 2020.

Dudzik, M.; Romanska-Zapala, A.; Bomberg, M. Ein neuronales Netz zur Überwachung und Charakterisierung von Gebäuden mit Umweltqualitätsmanagement, Teil 1: Verifizierung unter stationären Bedingungen. *Energien.* 2020, *13*, 3469.

Dudzik, M.; Toward characterization of indoor environment in smart buildings; Part 1: Verwendung des Kriteriums der vorhergesagten mittleren Stimmabgabe. *Nachhaltigkeit.* 220(12), 6749.

Romańska-Zapała, A.; Furtak, M.; Fedorczak-Cisak, M.; Dechnik, M., The Need for Automatic Bypass Control to Improve the Energy Efficiency of a Building Through the Cooperation of a Horizontal Ground Heat Exchanger with a Ventilation Unit During Transitional Seasons: A Case Study", *WMCAUS 2018*, Prague *IOP Conference Series: Materials Science and Eng.*, Vol. 246

Fedorczak-Cisak, M.; Bomberg, M.; Yarbrough, D.W.; Lingo, L.E.; Romanska-Zapala, A. Position Paper Introducing a Sustainable, Universal Approach to Retrofitting Residential Buildings. *Gebäude* . 2022, *12*(6), 46. https://doi.org/10.3390/buildings12060846.

Bomberg, M.; Kisilewicz, T.; Nowak, K. Gibt es einen optimalen Bereich der Luftdichtheit für ein Gebäude? *J. Build. Phys.* 2016, *39*, 395–420.

Klingenberg; Kernagis, K.M.; James, M. Homes for a Changing Climate: Passive Houses in the U. S. Paperback – 1. Januar 2009, Bestellung über das Internet

Wright, G.; Klingenberg, K. *Climate-Specific Passive Building Standards*; U. S. Department of Energy, Building America, Office of Energy Efficiency and Renewable Energy, Report, 2015.

Heibati, R.S.; Maref, W.; Saber, H.H.. Assessing the Energy and Indoor Air Quality Performance for a Three-Story Building Using an Integrated Model, Part 1: The Need for Integration. *Energies*. 2019, *12*(24), 4775.

Heibati, R.S.; Maref, W.; Saber, H.H.. Assessing Energy, Indoor Air Quality, and Moisture Performance for a Three-Story Building Using an Integrated Model, Part Three: Entwicklung des integrierten Modells und Anwendungen. Energies. 2021, *14*(18), 5648.

Hu, X.; Shi, X.; Bomberg, M. Radiant heating/cooling on interior walls for thermal upgrade of existing residential buildings in China. In Proceedings of the In-Build Conference, Krakau TU, Krakau, Polen, 17. Juli 2013, S. 314.

Kosuke, S.; Kataoka, E.; Horikawa, S.. Thermo-Active Building System Creates Comfort, Energy Efficiency. *J. ASHRAE*. March 2020, *62*(3), 42–50. ASHRAE.org.

Häupl, P.; Grunewald, J.; Fechner, H. Feuchteverhalten eines Gründerzeithauses durch eine kapillaraktive Innendämmung. In Proceedings of the Building Physics in the Nordic Countries, Göteborg, Schweden, 24–26 August 1999; S. 225–232

simonson, C.J.; Salonvaara, M.; Ojanen, T.. Heat and mass transfer between indoor air and a permeable and hygroscopic building envelope: P. II, Verifizierung und numerische Studien. *J. Bldg. Phys*. 2004, 2004, *28*, 161–185.

Bomberg, M.; Pazera, M. Methoden zur Überprüfung der Zuverlässigkeit von Materialeigenschaften für die Verwendung von Modellen in der hygrothermischen Echtzeitanalyse. In Res. in Building Physics-Proc.1st Central Eur. Symp. Building Physics (eds Gawin and Kisielewicz), Krakau-Lodz, Polen, 13–15 September 2010. 2010, pp. 89–107.

Vereecken, E.; Roels, S. Kapillaraktive Innendämmung: wiegen die Vorteile mögliche Nachteile wirklich auf? *Mater. Struct*. 2015 2015, 48(9), 3009–3021.

Fort, J.; Kocí, J.; Pokorný; Podolka, L.; Kraus, M.; Cerný, R.. Characterization of Responsive Plasters for Passive Moisture and Temperature Control. *Appl. Sci*. 2020.

De Deer, R.; Zhang, F. Dynamic environnent, adaptive comfort, and cognitive performance. In Proceedings of the 7th International Building Physics Conference, IBPC2018, Syracuse, NY, USA, 23–26 September 2018; pp. 1–6

Buratti, C.; Vergoni, M.; Palladino, D. Thermal comfort evaluation within non-residential environments: development of Artificial Neural Network by using the adaptive approach data. In 6th Int. Building Physics Conf., IBPC 2015, Energy Procedia, 2015, *78*, S. 2875–2880

Ferreira, P.; Silva, S.; Ruano, A.; Negrier, A.; Conceição; Eusébio Neura Network PMV Estimation for Model-Based Predictive Control of HVAC Systems. In WCCI 2012 IEEE World Congr. on Comp. Intelligence, Brisbane, Australien 15–22, 2012.

Private Gespräche mit IBACOS- und BCI-Teams der BA, 2018.

Fadiejev, J.; Simonson, R.; Kurnit ski, J.; Bomberg, M. Thermal mass, and energy recovery utilization for peak load reduction. *Energy Procedia*. 2017, 132, 38.

Bomberg, M. Anna Romanska-Zapala und David Yarbrough Auf dem Weg zu einem neuen Paradigma für die Bauwissenschaft (Bauphysik). *World* . 2021, *2*(2), 194–215. https://doi.org/10.3390/world2020013.

Covey, S.R. The 7 Habits of Highly Effective People. Simon, & Shuster, 1989, pp. 95–182.

IDP. An Integrated Approach to Design of Protocol Specifications Using Protocol Validation and Synthesis. *IEEE Trans. Comput*. Apr 1991, 459–467, 40. 10.1109/12.88465.

Bomberg, M.; Yarbrough, D.; Furtak, M. Buildings with environmental quality management (EQM), part 1: Entwurf einer multifunktionalen Baumatte. *J. Build. Phys*. . 2017, *41*, 193–208.

Bomberg, M.; Ein Konzept der kapillaraktiven, dynamischen Isolierung, integriert mit Heizung, Kühlung und Lüftung, Klimatisierung. Front. *Archit. Civ. Eng. China.* 2010, 4, 431–437.

Romanska-Zapala, A.; Bomberg, M.; Fedorczak-Cisak, M.; Furtak, M.; Yarbrough, D.; Dechnik, M. Buildings with Environmental Quality Management (EQM), part 2: Integration of hydronic heating/cooling with thermal mass. *J. Build. Phys.* 2018, *41*, 397–417.

Yarbrough, W.; Bomberg, M.; Romanska-Zapala, A.. Buildings with Environm. Quality Management (EQM), Teil 3: Vom Blockhaus zum Null-Energie-Gebäude. *J. Build. Phys.* 2018.

Romanska-Zapala, A.; Bomberg, M.; Yarbrough, D. Buildings with Environmental Quality Management (EQM), part 4: A path to the future NZEB. *J. Build. Phys.* 2018, *43*, 3–21.

Torrie, R.; Bak, C. Building Back Better with a green renovation wave, (Planung für eine grüne Erholung), Internet-Newsletter, 22. April 2022 (eigenes Archiv), 2022.

Meadows, D.. The limits of growth, The Donella Meadows project, Academy of System Change, im Internet; siehe auch "Meadows on social paradigms Meadows, D. H.; Meadows, D.L.; Randers, J.; Behrens III, W.W. The Limits to Growth; Universe Books: NY, NY, USA, 1972

# Executive summary in Polish
## Termomodernizacja w świetle strategicznych poglądów na energie i środowisko mieszkaniowe rozszerzone streszczenie książki

Przez dwie dekady, kilku naukowców współpracujących w kontakcie wirtualnym, pracowało nad integracją monitorowania i modelowania sprawności technologii rehabilitacji istniejących mieszkań i budynków (Brennan et al., 2008, Bomberg et al., 2009, Walburger et al., 2010, Mattock, 2010, Thorsell and Bomberg, 2011, Hu and Shi 2013, Kloseiko and Kalamees, 2018, Heibatti et al., 2019. Yarbrough et al., 2021). Przewidywanie rozwoju technologii budowlanej z równoczesną oceną jej sprawności nie jest łatwe. Taka ocena dotyczy ogrzewania w zimie, chłodzenia w lecie, wentylacji dla osób zdrowych, ale wentylacja dla osób chorych (na grypę czy covid 19), musi uwzględnić wpływ lokalizacji mieszkania w porównaniu z płaszczyzna neutralna ciśnienia powietrza w budynku. Projektant musi również zaplanować jakie urządzenia można umieścić w ścianach budynku oraz znaleźć miejsce na zbiorniki wody, które zastąpią pojemność cieplną gruntu (Bomberg et al., 2020). Technologia przedstawiona w tej książce, zaczyna się tam, gdzie obecna technologia domów pasywnych kończy się (Klingenberg et al., 2009, Wright and Klingenberg, 2015). Dlatego przyjęliśmy angielską nazwę PTAC (Passive and Thermo-Active Cluster) tłumacząc na Polski: PTAK (pasywny i termo-aktywny kompleks). Technologia PTAK wnosi: (1) dwustopniowy proces budowy, (2) po angielsku MAPE (monitoring and performance evaluation), w tłumaczeniu MMSB (Monitorowanie i Modelowanie Sprawności Budynku), (3) integracje systemu hydraulicznego, (4) z nową metodą dostarczania energii oraz (5) system kontroli automatyki budowanej.

W tej książce demonstrujemy system PTAK zapewniający 90 procent zmniejszenia energii w nowym budownictwie oraz 70 do 80 procent redukcji energii w porównaniu z poziomem referencji jakim są wymagania państwowych czy Europejskich standardów w roku 2004 (Yarbrough et al., 2018). Ponadto postulujemy, że osiągniemy to przy dzisiejszym koszcie tzw. głębokiej renowacji.

Jednakże, nawet najlepsza technologia sama w sobie, nie rozwiązuje problemu renowacji istniejących budynków. *Obecny sposób traktowania termomodernizacji wymaga zasadniczych zmian strukturalnych.* Publiczno-prywatne projekty demonstracyjne są środkiem dla rozwiązania problemu i demonstracji tego rozwiązania. *Takie projekty mogą zapoczątkować fale informacji publicznej niezbędnej dla nowej roli termomodernizacji, a mianowicie zwolnienia tempa zmiany klimatu (Fedorczak-Cisak et al. 2022)*

https://doi.org/10.1515/9783112217023-204

## Potrzeba wizji dla termomodernizacji

Covey (1989) w swojej znakomitej książce, mówi zacznij każdy projekt z myślą o skończeniu, co chcesz osiągać. Az do lat 1990tych projektowaliśmy budynki w oparciu o tradycje. Dopiero wtedy, kiedy społeczeństwo zaczęło wymagać więcej niż umieliśmy, dodano modelowanie energii. To stworzyło zintegrowany zespół projektowy, po angielsku „Integrated Design Protocol" (IDP, 1991). Ten zespól już w projekcie koncepcyjnym zajął się sprawami energii środowiska wewnętrznego.

Patrząc na kraje rozwinięte gospodarczo, widzimy cały szereg świetnych rozwiązań, ale żadne z nich nie jest uniwersalne (Fadiejev et al, 2019, Kloseiko and Kalamees, 2016, 2107).

Budynki pasywne w Niemczech nie są podobne do budynków a USA; budynki w Teksas nie są podobne do budynków w Illinois. Jednakże, patrząc na historie (Bomberg and Onysko, 2008), i porównując ja z rozwojem samochodów. ponieważ budownictwo mieszkaniowe jest teraz na granicy rozwoju przemysłowego (Chitwood and Harriman, 2011), widzimy ze rewolucja techniczna spowodowana była zmianą sposobu myślenia. Obecne koszty budowy, podniosły się po okresie pandemii tak wysoko, ze aby uzyskać przestępną dla ludzi technologie budowlaną, musimy zażądać, aby ona była zależną od klimatu ale uniwersalną (Bomberg et al., 2017). *A zatem, nasza praca zaczyna się od wizji.*

Potrzebę klimatu-zależnej, uniwersalnej, technologii renowacyjnej zauważyliśmy w roku 2010 (Bomberg et al., 2010), w czasie Zachód-Chiny konferencji fizyki budowlanej w Nanjing, gdyż żadna z ówczesnych metod renowacji nie nadawała się do klimatu w tym rejonie Chin, zimnych zim i gorących lat, przy wysokiej wilgotności powietrza przez cały rok. Obniżenie wilgotności nie jest problemem, wtedy, kiedy powietrze jest cyrkulowane wewnątrz budynku. Jednakże, przez wiele lat panowała stagnacji nauk budowlanych w Ameryce (Bomberg et al., 2020, 2021),i uproszczony dla celów 'zielonej polityki w Europie" zamiast wentylacji i recyrkulacji powietrza koncept otwierania okien. To ostatnie nie zmniejsza poziomu zanieczyszczeń bakteriologicznych *i doświadczenia z epidemii SAR-coV2 uczą ze redystrybucja i filtracja powietrza muszą być normalną praktyką budowlaną.*

Woda, mając cztery razy większą pojemność ciepła niż powietrze, jest wybranym nośnikiem energii grzewczej, a więc ogrzewanie i chłodzenie musi być oparte na nisko-temperaturowych obwodach hydraulicznych. Mamy wiec, dwa niezależne systemy kształtujące środowiska wewnętrzne. Wizja termomodernizacji musi zawierać integracje systemową tak aby stworzyć uniwersalną technikę do stosowania w różnych klimatach. W tej książce pokazujemy nawet ze w przyszłości gospodarka woda musi być włączona do sytemu ogrzewania i chłodzenia (Bomberg et al., 2020).

Poza Ameryka, gdzie dzisiaj nauki budowlane staja się jednym ze specjalizacji zawodowych, ale nie są prowadzone na uczelniach; na uczelniach Europejskich to jest marginesowy przedmiot bez wpływu na rynek. W efekcie rządy we wielu krajach uprawiają politykę zielonych grantów, i. e., sponsorują „zielone produkty", które często maja negatywny efekt na rozwój gospodarki. Dla przykładu. poprzedni rząd dawał granty dla form produkujących pompy ciepła, a jednocześnie dla ludzi kupujących pompy ciepła. Jedyne co osiągnięto to podniesienie się cen pomp ciepła oraz negatywna reakcje społeczeństwa, gdyż wiele z tych „sponsorowanych okazji" było na bardzo niskim poziomie technicznym, np. najpopularniejszy sponsorowany zbiornik gorącej wody, dostarczał 81% energii było z grzejnika elektrycznego. (W tej książce tłumaczymy, dlaczego pompa do zbiornika buforowego nie może być tą samą, która dostarcza gorącą wodę użytkową). Gdyby istniała wizja termomodernizacji nie byłoby tak nieudolnych akcji rządowych. Nie można wymagać wiedzy od polityków, ale też nie wolno używać polityki do bezsensownego manipulowania rynku.

## Koncepcja Pasywnego i Termo-Aktywnego Kompleksu (PTAK)

Ten przegląd zaczynamy od niepublikowanego raportu Państwowej Rady Badawczej Kanady (raport 1639, Grudzień 1947) który przedstawił wyniki badan dwu, jednopiętrowych budynków. Dodatkowa płyta piwniczna z metalowymi rurkami grzewczymi, była umieszczony pod 1.2 m wysokim podpiwniczeniem. Strat ciepła dla zadanego poziomu wentylacji przedstawione były w funkcji temperatury zewnętrznej.

Sześćdziesiąt lat później wynalazca węgierski zastosował taki sam system z rurkami PEX (zbrojony polietylen) i pompa wody dostarczająca energie geotermiczna do ogrzewania podłogowego, nazywając to aktywną izolacją ciepła (Kisilewicz et al. 2019, 2023). Podobnie do węgierskiego, nagrodzony w Ameryce, system termo-aktywny w budynku w Tokio, używał jednocześnie wodny system grzewczo-chłodzący i wentylacyjny (Kosuke et al. 2020). W tym samym czasie, wiedząc ze wiele detali musi być zintegrowanych w budynku (Mattock, 2919), zespól projektowy w Montrealu wprowadził wielostopniowy proces budowy zmniejszając zużycie energii poniżej 10% stanu początkowego. To wymaganie było opublikowane w roku 2008 przez Lawrence Berkeley National Laboratory (LBNL, Bomberg and Onysko 2008).

Jeżeli zrobimy syntezę tych czterech organizacji: Atelier Rosemount (Information notes, 2016), Montreal, Kanada (Heibatti, 2019, 2021), węgierski wynalazca, (Kisilewicz et al. 2019, 2023), Tokio zespól projektowy (Kosuke et al. 2020). i Państwowe Laboratorium LNBL w Ameryce, to możemy stwierdzić *ze technologia nowych budynków ekologicznych z blisko zera emisją węglową i energią była już wypracowana w roku 2020.*

Pozostały budynki istniejące, gdzie dostęp do paneli słonecznych był utrudniony. Jednakże, rozbudowując zintegrowany system poprawy środowiska wewnętrznego i dodając wentylacje oraz gospodarkę woda użytkową znajdujemy dobre rozwiązanie dla istniejących budynków (Bomberg et al. 2016a, Romańska-Zapala 2018). Ekonomia gospodarki woda wymaga zbiorników wody, które mogą również służyć jako dolne źródło pompy ciepła. Budynek doświadczalny pracy doktorskiej w Syrakuzach, NY (Lingo and Roy, 2016) miał duży, podziemny zbiornik wody i pozwalał na badanie rożnych systemów grzewczych. Tak więc słowo kompleks, w nazwie technologii PTAK oznacza albo kombinacje różnych metod technicznych albo grupę budynków, albo budynek z otoczeniem gruntowym. To jest bardzo ważna zmiana, gdyż wprowadza możliwość klimatycznego polaczenia sieci rożnych budynków (patrz tekst o rejonowej sieci klimatycznej).

PTAK jest uniwersalna, klimat zależna, publicznie dostępną technologią przygotowana jako punkt początkowy dla różnych opcji, np. sieci neuronowe (Romańska-Zapala and Bomberg, 2019, Dudek et al., 2020, Dudzik et al. 2020, 2021) mogą współdziałać z komercjalnymi programami do usług i komfortu. Wymaganie kompleksu pozwala na umieszczenie zbiornika wody w gruncie. Jeśli nie jest to możliwe, zbiorniki wody maga być umieszczone na sianach łazienki, klatki schodowej, w piwnicy a nawet w szopie na zewnątrz budynku.

Wielostopniowy projekt w Montrealu, Kanada przełamał barierę dostępności wprowadzać krótkie okresy pożyczek hipotecznych dostępnych dla finansowania renowacji budynku przez mieszkańców. Pokażemy to na przykładzie. Załóżmy ze mieszkaniec może płacić 4 tysiące (walutą kraju) przez dwa lata 2.5tysiecy przez 4 następne lata to jest 18 tysięcy przez 6 lat, obecnie płaci 330 za ogrzewanie a przy inwestycji 30 tysięcy będzie płacił tylko 150. Po następnych 20 tysiącach będzie płacił zero. Mieszkaniec bierze pożyczkę 30 tys. na 6% i płaci 705 miesięcznie zamiast 330. Po 3 latach spłacił 21,868 z funduszu i 3,768 z kaszty bankowe. Wtedy bierze druga pożyczkę na 30 tysięcy (ma 8,1 tys. z pierwszej pożyczki do spłacenia). W sumie przez 6 lat zapłacił 7500 na koszty bankowe, 44000 na spłacenie długu z banku, 6000 własnych funduszy oraz 12000 na koszty ogrzewania.

Jeśli nie ma dodatkowych pieniędzy to ten sam przykład wymaga dłuższego czasu na spłacania. Obecnie wiele banków wydaje pożyczki na podstawie zainwestowanego kapitału. Pożyczka składa się z kapitału, oprocentowania, podatku i ubezpieczenia. Jeżeli dodamy koszt energii to podniesiemy wartość pożyczki nie wpływając na koszt życia. Jednak inwestycja tego typu jest dobra dla społeczeństwa (obniża zużycie energii), lokalnej ekonomii (prace remontowe) oraz mieszkańca (lepszy klimat wewnętrzny, Bomberg et al., 2021).

## Opis systemu PTAK (pasywny i termo-aktywny kompleks)

Koszt termomodernizacji jest ponad 50% wyższy niż koszt budowy do minimalnego standardu. A więc, pierwszy problem do rozwiązania jest zmiana obecnego procesu budowlanego na proces dwu lub wielostopniowy. Drugi stopień procesu zaczyna się jakiś czas później, kiedy budynek wybudowany jest do stopnia podnoszącego kredyt bankowy. Oczywiście projekt jest zrobiony na wszystkie fazy budowlane i pozwala na poniesienie kredytu. Poniżej, podajemy przykład takiej realizacji przy budowaniu osiedla nazwanego Atelier Rosemont w Montrealu, Kanada. Konstrukcja rozpoczęta była w roku 2008 a skończona w 2018, kiedy uzyskano 92% redukcji początkowej energii.

Proces renowacji wykonany był w następujących stopniach:
- Poprawa izolacji ścian zewnętrznych, woda użytkowa, termiczne panele słoneczne dały 36% początkowej redukcji,

–   Szara woda oraz wszystkie pasywne metody = 42% redukcji
–   Ogrzewanie podłogowe z pompa ciepła = 60% redukcji
–   Gorąca woda użytkowa z paneli słonecznych = 74% redukcji
–   Photovoltaiczne panele słoneczne podniosły całkowitą energię do 92% redukcji

Powyższe cyfry zostały wielokrotnie potwierdzone w programie Building Ameryka (private communications, 2018)

W czasie wykonywania naszego projektu ustaliliśmy ze ustawienie wymienników ciepła w ścianach zamiast w podłodze wprowadza zmniejszenie energii dla uzyskania tego samego efektu. Czasie ogrzewania mieszkania w Nanning, China (Hu, 2013) zamiast 98 GJ potrzeba tylko 58 GJ a w czasie chłodzenia zamiast 31 GJ tylko 24 GJ. Panele w tym projekcie były wykonane z cementem magnezowym i zawierały zmielone włókno drewniane oraz krzaków ryżu, aby podnieść pochłaniane wilgoci i elastyczność tynku.

Następny problem jest z wyborem pompy ciepła. Tradycyjny wybór w USA czy w Europe jest powietrze-woda z niskim współczynnikiem sprawności 2.5 do 3 podczas gdy my wymagamy typ woda-woda aby osiągnąć współczynnik sprawności ponad 4. Dlatego w technologii PTAK zamieniliśmy geotermiczny wymiennik ciepła na dwa zbiorniki wody oraz niezależną pompę wody (Bomberg et al., 2015). Zastosowanie pompy ciepła woda-woda ma szereg korzyści, a namówicie istnienie cieplej i zimnej wody przez cały rok, łatwe wprowadzenie gospodarki wodą szarą, łatwa integracja chłodzonych paneli słonecznych, łatwe dodawane mocy pompy ciepła, jak również innowacja rejonowej sieci klimatycznej, która także nadaje się do budynków historycznych.

Tym niemniej, sama technologia, nawet najlepsza, ale bez ustalonych wymagań społecznych nie rozwiążę problemu zmiany klimatu. Kanadyjska demonstracja technologii budynków pasywnych nie została przyjęta przez rynek (Regina Conservation House, 1978), gdyż rozdział pomiędzy naukami budowalnymi a praktyką budowlaną był zbyt duży. Kanada i Ameryka stworzyły publiczno-prywatne projekty demonstracyjne, (R2000 Standardy, 2000, and Building Ameryka), które zamieniły paradygmat myślowy z poprawy materiałów stosownych do budowy na projekt całego budynku, w którym materiały wybrane są tak aby uzyskać wymaganą sprawność budynku.

Kanada i USA wzorowały się na rewolucji technicznej przemysłu samochodowego. Henryk Ford zamienił paradygmat myślowy, kiedy zamiast pracownika podchodzącego do budowanego samochodu, samochód podjeżdżał do pracownika. Podobna rewolucja techniczna (Kuhn, 1970) jest potrzebna przy rekonstrukcji istniejących budynków. Taka rewolucja ma wprowadzić wygrana dla społeczeństwa, ekonomii oraz mieszkańca budynku. Społeczeństwo wygrywa przez zwolnienie szybkości zmian klimatu, ekonomia zyskuje nowe, lokalne miejsca pracy, a mieszkaniec przez zmniejszenie kosztów ogrzewania i poprawę jakości środowiska wewnętrznego. Pamiętając, że budowniczy zrobi tylko to, za co ktoś zapłaci, społeczeństwo musi wymagać budynków o zerowej emisji i poprawy klimatyzacji budynków.

Proponowana metoda PTAK (Pasywny oraz Termo-Aktywny Kompleks) wprowadza dla grupy budynków:

– Dwustopniową (wiele-stopniową) konstrukcje, która zmienia proces finansowania

– Automatykę budowlana dla kontroli *adaptacyjnego* klimatu wewnętrznego (De Deer and Zhang, 2018) jak również udziału masy cieplnej oraz integracji paneli słonecznych z pompą ciepła.

– Monitorowanie i modelowanie sprawności budynku (MMSB), dla kalibracji modelowania energetycznego i cieplno-wilgotnościowego, optymalizacji zużycia energii oraz poprawy środowiska wewnętrznego podczas użytkowania.

– Budynek z otoczeniem lub grupę budynków dla umieszczenia zbiorników wody oraz wprowadzenia rejonowej sieci klimatycznej.

Szereg elementów metody PTAK było już zastosowane w budynkach w Ameryce, w Kanadzie, na Węgrzech oraz w Japonii. Budynek termo-aktywny w Tokio, Japan zdobył główną nagrodę w ASHRAE (American Society of Heating, Refrigeration and Air Conditioning Engineers) w roku 2020. Przedstawiając w tej książce syntezę kilku technologii, stawiamy pytanie czy możemy również stworzyć publiczno-prywatny projekt dla demonstracji technologii PTAK

## Logika wyboru systemu energetycznego

Pojemność cieplna wody jest ponad cztery razy większa niż powietrza, a więc analizujemy tylko systemy hydrauliczne z zastosowaniem pompy ciepła woda-woda. Technologia musi być uniwersalna z modyfikacjami zależnymi od typu budynku oraz klimatu, aby z czasem zmniejszyć koszty budowy. Dlatego szukamy tylko takich metod, które można wielokrotnie powtórzyć.

Analizę rozpoczynamy od niepublikowanych badan Państwowej Rady Badan Naukowych Kanady (Raport 1639, Grudzień 1947). Ilość ciepłą dostarczanego do płyty grzewczej (o wymiarach równych z podłoga budynku) przedstawiona była w funkcji temperatury zewnętrznej oraz zastosowanej wentylacji mechanicznej (od 1.5 do 4 wymian powietrza na godzinę). 60 lat później, węgierski wynalazca, zastosował identyczny układ z betonowa płytą grzewcza umieszczona pod pasem ziemi, na której spoczywa strop piwnicy i która to płyta pracuje jako wymiennik ciepła geotermicznego. Ten system nazwano termicznie aktywna izolacja (patrz Kisielewicz et al., 2019, 2023).

W Montrealu, Kanada, dla ułatwienia integracji wielu małych osiągnięć, zastosowano wielostopniowy proces budowy który trwał od 2008 do 2018 i zredukował zużycie energii o 92 % od poziomu wymagań kodu budowlanego 2004. To odpowiada wymaganiom Państwowego Laboratorium Lawrence Berkeley (patrz Amerykańska konferencja

BEST1, Bomberg and Onysko, 2008). Tak więc, w roku 2020, robiąc syntezę prac w Montrealu, na Węgrzech, w Japonii oraz obliczeń w Ameryce, mamy *metodę budowania nowych budynków zrównoważonych z zerem emisji dwutlenku węgla i blisko zerowej energii użytkowej.*

Do rozwiązania (Bomberg et al., 2017, Yarbrough et al., 2019, Heibatti et al. 2019) pozostaje tylko renowacja istniejących budynków, gdyż wciąż technologia budynków pasywnych przy zastosowaniu pompy ciepła do hydraulicznego ogrzewania dostarcza 55– 60 % redukcji energii (zakres odnosi się do różnych klimatów która były badane w USA). Będziemy przedstawiali praktyczne rozwiązanie w tej książce na 10 do 20 % zwiększania redukcji energii tak aby zastąpić brak paneli słonecznych i otrzymać około 70 % redukcji niezbędnej do wprowadzania rewolucji naukowej w modernizacji budynków.

Rozwiązanie tego problemu wymaga rozszerzenia stopnia integracji systemu. Lekcje z pandemii SAR-coV2 (covid 19) mówią ze filtracja powietrza jest niezbędną, a więc potrzebna jakaś siła napędowa dla wentylacji, najlepiej w formie hybrydowej. Szczeliny powietrzne pomiędzy istniejącą ścianą a panelami grzewczo-chłodzącymi służą do dystrybucji. Doświadczenia z Kalifornii wykazują ze musimy także uwzględnić zmienną szybkość wentylacji pomieszczeń.

Drugi element rozszerzenia zakresu pracy nie jest oczywisty. Cena energii geotermicznej jest zmienna i trudno jest ją obniżyć. Możemy jednak, liczyć na podniesienie sprawności pompy ciepła woda-woda, jeśli dolne źródło ciepła jest bliższe do nominalnej temperatury gorącej wody użytkowej. Ponadto doświadczenie mówi ze około 1/ 3 energii z prysznica można odzyskać, jeśli wprowadzimy gospodarkę wodą szarą, a więc zmieniamy paradygmat myślenia, aby zacząć mówić o budynku razem z jego otoczeniem lub o grupie budynków. Duzy wkład w role gruntu dla wstępnej klimatyzacji powietrza wentylacyjnego zrobiła Politechnika Krakowska (Flaga-Maryanczyk et al., 2014, Romańska-Zapała et al., 2018, Yarbrough et al., 2019, Fedyczak-Cisak et. al., 2022)

W języku Angielskim mówimy o grupie budynków czy metod pomiarowych używając tego samego słowa „cluster", po Polsku wybraliśmy słowo „kompleks". W efekcie w systemie PTAK nie ma różnicy pomiędzy budyniem istniejącym czy nowo projektowanym, budynku mieszkalnym czy historycznym, jeśli on dzieli grunt z innym budynkiem. Nie ma różnicy pomiędzy jednym budynkiem z otoczeniem, czy setka budynków na osiedlu.

PTAK jest uniwersalną, klimatycznie przystosowaną, publicznie dostępną technologią, w której są różne możliwości wyboru rozwiązań technicznych. Projektant może wybrać wielkość pompy ciepła i zbiorników wody, używać pompę ciepła tylko w nocy lub przez cały czas, wybrać wyższą lub niższą temperaturę wody gorącej, wybrać frakcje ścian pokrytych wymiennikami ciepła itd. Tym niemniej, jeżeli ta technologia znajdzie duże zainteresowanie, zwiększy się ilość usprawnień i obniży się jej koszt stosowania.

## Termomodernizacja może zmniejszyć koszt emisji węglowej

Torrie i Bak (2022) pisząc o Kanadzie, podkreślili ze pomimo piętnastu milionów domków oraz mieszkań, z 2.1 bilionami metrów kwadratowych oraz 65 milionów ton emisji węglowej (2/3 to gaz) Kanada nie ma drogi rozwiązania problemu i porównali sytuacje do Amerykańskiej rodziny, która w roku 1910 chce kupić samochód. Używając dzisiejszej wartości, samochód kosztował 25 tysięcy dolarów a Ford sprzedał wtedy 19 tysięcy na rok. 10 lat później Model T kosztował 5 tysięcy a Ford sprzedał 941 tysięcy.

60 bilion dolarów wydane było w Kanadzie w roku 2019 na renowacje budynków a tym polowa na ogrzewanie stosując pompy ciepła. Przeciętny koszt renowacji wynosił 40 tysięcy dolarów, co przy 10 latach przejścia na nowe technologie daje koszt 100 amerykańskich dolarów za 1 tonę śladu węglowego.

Gdyby można było teraz wprowadzić technologie która obniży koszt o 33 % to by podniosło ilość renowacji trzykrotnie to koszt 1 tony CO2 zszedłby do 15.6 dolara. Ponadto, jeśli ta technologia ma również 30% mniejszy ślad węglowy to koszt 1 tony CO2 spadnie poniżej 10 dolarów amerykańskich.

Tak więc mamy wyznaczony cel do osiągniecia. *Mamy podnieść efektywność nowej technologii o 1/3 przy tej samej cenie, kiedy wprowadzimy uniwersalna i powszechnie dostępną, publiczna technologie PTAK, aby zmniejszyć koszt 1 tony CO2 poniżej 10 dolarów amerykańskich.* To pozwoli nam dopiąć celu zmniejszenia zmian klimatu poprzez termomodernizacje budynków.

## Wielo-stopniowe budowanie może zmniejszyć koszt jakości budynku

Rozdział pomiędzy nauka i praktyką (udokumentowana w Kanadzie dwukrotnie (Bomberg et al., 2009) w prowincji Saskatchewan, oraz w Atelier Rosemont w Montrealu. Taki brak postępu na rynku, spowodował stworzenie prywatno-publicznych programów demonstracyjnych, które obejmowały wkład edukacji publicznej. Te programy zmieniły zrozumienie publiczne wymagań technicznych dla budynków mieszkalnych i autorzy są przekonani ze należy takie programy powtórzyć w tych krajach które chcą uniknąć recesji spowodowanej obecnymi wojnami handlowymi. Chcemy także podkreślić, że Atelier Rosemount w Montrealu, przełamał barierę nadmiernych kosztów. W osiedlu, gdzie luksusowe mieszkania ponad 100 $m^2$ z oknami na dwie strony świata dla uzyskania wentylacji naturalnej, miały za sąsiadów mniejsze mieszkania socjalne, z podobnym komfortem mieszkań, wszystkie koszty były pokryte przez oszczędności własne lokatorów oraz wielokrotne powtarzanie pożyczek hipotecznych.

Wiele problemów w budownictwie spowodowanych jest dzisiaj przez nieprzemyślane decyzje administracyjne wynikające z braku wkładu naukowego. Akademickie grupy fizyki budowlanej Europie nie uwzględniają ani praktyki i kosztów operacyjnych, a więc nie maja wpływu na rozwój budownictwa. Fizyka budowlana nie jest normalnie uczona przez uczelnie w USA. Było szereg prób zmiany, ale bez efektu. Państwowa Rada Badan Naukowych w Kanadzie w latach 1960–1980 próbowała poprawić system nauczania, Państwowy Instytut Standardów, Washington, USA, (1970–1980) próbował wprowadzić metodologię analizy funkcjonalnej do budownictwa. Można sądzić, że głównym powodem niepowodzenia były ASTM (Amerykańskiego Stowarzyszenia Badan i Materiałów) metody badawcze które są oparte na porównaniu materiałów.

W nowej generacji „zrównoważonego środowiska budowlanego" potrzebujemy stosować integracje badan i modelowania. Czy są to modele funkcjonalne, numeryczne czy też oparte na sieciach neuronowych, modele musza być porównane (kalibrowane) z danymi zmierzonymi na rzeczywistych budynkach (Klõšeiko and Kalamees, 2016, 2018, Dudek et al., 2020, Dudzik, 2020, Dudzik et al., 2020, Heibatti et al., 2019, 2021). Ponadto każdy przepływ ciepła powoduje przepływ wilgoci, a więc nie możemy pominąć spraw wilgotnościowych (Häupl et al., 1999, Simonson et al., 2004, Bomberg M and Pazera, 2010, Vereecken and Roels. 2015, Fort et al., 2020), A więc, po angielsku „Monitoring And Performance Evaluation" (MAPE), czy po Polsku „Monitorowanie i Modelowanie Sprawności Budynku (MMSB) jest centralnym elementem nowej fizyki budowlanej.

Nadepną sprawą jest kryterium sprawności. Powinniśmy stosować metodologie stanów granicznych do zagadnień trwałości budynku czy inżynierii środowiska. Należy się jednak spodziewać ze komisja kwalifikacyjna potwierdzająca curriculum będzie wymagać ujednolicony typ konstrukcji dla analizy renowacji ścian i proponujemy, aby początkowo był to system PTAK. Później, CESBP (Sympozjum Fizyki Budowlanej w Centralnej Europie) może przeląc wypracowanie systemu, dla którego analizę wprowadzimy do uniwersyteckich kursów fiyki budowli.

## Rejonowa siec klimatyczna

Technologia PTAK pozwala na wykorzystanie cieplej wody powracającej ze ściennych wymienników ciepła w dwóch różnych możliwościach: (1) w układzie jednego domu, powraca ona do zbiornika wody zimnej, aby zostać podgrzana, lub (2) w układzie rejonowej sieci klimatycznej, jest ona wysłana do następnego budynku w sieci. Uniwersalny charakter technologii PTAK pozwala na ten wybór niwelując różnicę pomiędzy pojedynczym budynkiem czy budynkiem w grupie osiedlowej. Ta innowacja rejonowej sieci klimatycznej (RSK) może także być zastosowana do budynków historycznych umożliwiając umieszczenie wszelkich urządzeń mechanicznych w sąsiednim budynku i traktując dwa budynki jak jedna całość z kilkoma strefami klimatycznymi. Lokalna siec klimatyczna doprowadza również powietrze wentylacyjne o temperaturze bliskiej temperatury gruntu.

Ponieważ technologia PTAK wprowadza także zbiorniki wody pełniące funkcje magazynu ciepła, linia lokalnej sieci klimatycznej włącznie ze zbiornikami wody może być albo częścią kompleksu budynkowego lub częścią system dostawy wody miejskiej. Podsumowując, ponieważ koszt konstrukcji grupy budynków jest niższy niż koszt budowy tych budynków oddzielnie, dodając RSK do projektu osiedla zmniejsza koszt operacyjny tych budynków.

## Perspektywa technologii PTAK

Odwiedzając grupę fizyki budowalnej w Montrealu zauważyłem ze nikt nie wiedział o osiedlu Atelier Rosemount, gdzie inwestycja była płacona przez mieszkańców i tylko z oszczędności energii osiągnęła poziom 92% przez 10 lat. Bez zaangażowania grupy naukowej, społeczeństwo nie wiedziało o tym osiągnieciu i nikt nie pociągnął tej inicjatywy dalej. Z kolei, brak zainwestowania wynika z nieudolności zawodowej ludzi, którzy chcą wprowadzić budownictwo ekologiczne, ale nie potrafią stworzyć warunków rynkowych dla zapotrzebowania rynkowego. Produktywność i zdrowie mieszkańców są głównymi kryteriami dla mieszkańców podczas gdy energia jest nieuchwytna. Tak więc, mówiąc tylko o jednym aspekcie, stwarzamy wrażenie ze technologia ekologiczna nie jest zrównoważona. Medows (1972) mówiła ze granty, zniżki podatkowe i podobne środki popierające ekologiczne budownictwo nie maja wpływu, i powinniśmy patrzeć na najwyższe społeczne wartości takie jak zmiana klimatu zamiast mówić o korzyściach finansowych.

Kiedy w latach 1980 i 1990 Kanada i USA wprowadziły narodowe programy publiczno-prywatne dla zmniejszenia energii nacisk był na całościowe podejście do budynku a nie tylko charakterystykę energetyczna. Każdy z zespołów uczestniczących miał wolny wybór technologii, a jeśli projekt wykazał również oszczędności energii to zespól dostawał dalszy grant. In kontrast do tych projektów, dzisiaj, typowy sponsorowany projekt e.g., Horyzont 2020 wydaje się być podstawa dla przyszłego rozwoju, ale następny projekt wymaga następnego przetargu. Poza Belgią i Chechią nie ma na świecie żadnej ciągłości w fizyce budowlanej.

Ten przegląd zaobserwował brak ciągłości pomiędzy nauką i praktyką budowlaną na przykładzie budynku kanadyjskiego budynku demonstracyjnego stosującego amerykańską technologie domów pasywnych (rok 1978). Równie dobrze moglibyśmy dać przykład z roku 2024, gdyż w północnych rejonach Europy w typowym rekuperatorze powietrza nie zastosowano podgrzewacza i woda zamarzała w wentylatorze. Brak ciągłości pomiędzy nauka i praktyka pozostaną, bo nie zmieni się struktura Europejskich uczelni, w Stanach Zjednoczonych zmieniła się tylko w przodujących uczelniach, pozostałe są tak samo zacofane jak większość Europejskich.

Dlatego nowa generacja technologii budowlanej musi być połączona z konfrontacja naszych czasów, a mianowicie zmiana klimatu. W takim przypadku, system kontroli środowiska musi spełnić trzy warunki: (1) pozwalać na równoległą regulacje automatyki budowlanej oraz mieszkańca, (2) być technicznie znakomitym, oraz (3) mieć umiarkowany koszt. Progres od kryzysu energetycznego w roku 1973 był znakomity (Klingenberg et al., 2009; Wright et al. 2015). Budynki pasywne pozwalają oszczędzi 40–45% energii a z użyciem pompy ciepła nawet 55–60%. Następne 10 do 20% daje metoda dostarczania energii proponowana w technologii PTAK (adaptacyjny klimat wewnętrzny).

Celem technologii PTAK jest rozszerzenie możliwości wymiany nakładów finansowych na krytyczne podzespoły, dostarczania czy magazynowania energii jak również sposobu jej użytkowania. Mattock (2010), który zaprojektował budynek drewniany z maximum bezwładności cieplnej wskazał na konieczność dalszej analizy pojemności cieplnej. W tym celu wprowadzamy dwa pojęcia magazynowania energii: na okres 24 godzin i na okres 7 dni. Fadiejew (2017) wskazał ze budynki betonowych z ociepleniem zewnętrznym maga dać wystarczającą pojemność cieplna na okres 24 godzin, w innych konstrukcjach system zbiorników wodnych i rurek PEX wypełnionych woda może zastąpić masę konstrukcji murowanej czy betonowej. W technologii PTAK współpracujemy z nowoczesna siecią elektryczna dostarczając i odbierając energie w czasie dogodnym dla sieci. W technologii PTAK nie analizujemy magazynowania energii elektrycznej, ponieważ w renowacji istniejących budynków magazynowanie energii elektrycznej jest zazwyczaj nie ekonomiczne.

Naszym problemem jest znalezienie drogi komunikowania się z ludźmi, którzy chcą podnieść swój standard mieszkaniowy. Wiemy teraz ze otwieranie okien nie jest efektywne dla usunięcia wirusów a filtracja powietrza w mieszkaniu najlepiej działa z indywidualnym wejściem świeżego powietrza. A więc, mamy wybór: czekać 20 lat aż naturalny rozwój technologii dojdzie do nowego poziomu wiedzy lub *stworzyć fale społeczną, aby zmienić system rekonstrukcji budynków*, natychmiast. To ostatnie rozwiązanie nazywa się rewolucją naukowa (Kuhn, 1970).

W technologii PTAK my proponujemy to drugie rozwiązanie i podkreślajmy trzy stopnie akcji w czasie: (1) projektowania, (2) monitorowania (kalibrując model do zmierzonych danych), zas (3) do optymalizacji systemów mechanicznych stosowanych w kontroli klimatu wewnętrznego. Ten proces wymaga modelowania produkcji, magazynowania i poprawy budowlanej.

# References

Bomberg, M.; Shirtcliffe, C.J. Influence of moisture and moisture gradients on heat transfer through porous building materials, Thermal Transmission Measurements of Insulation. In ASTM STP 660 (ed) R.P,;. Tye, ASTM, Philadelphia, 1978, pp. 211–233.

Bomberg, M.; Onysko, D. (Eds.) Energy Efficiency and Durability of Buildings at the Crossroads.2008. http://thebestconference.org/BEST1 (accessed on 25 February 2020).

Bomberg, M.; Brennan, T.; Henderson, H.; Stack, K.; Wallburger, A.; Zhang, J., 2009, High Environmental Performance (HEP), residential housing and building technology, for NY state. *A Final Report to NY State Energy Research and Development Agency and National Center of Energy Mgmt.*, manuscript, 2009, USA.

Bomberg, M.; Pazera, M. Methods to check reliability of material characteristics for use of models in real time hygrothermal analysis. In Res. in Building Physics – Proc.1st Central Eur. Symp. Building Physics (eds Gawin and Kisilewicz), Cracow–Lodz, Poland, 13–15 September 2010. 2010, pp. 89–107.

Bomberg, M.;. A concept of capillary active, dynamic insulation integrated with heating, cooling and ventilation, air conditioning system. *Front. Archit. Civ. Eng. China.* 2010 2010, 4, 431–437.

Bomberg, M.; Gibson, M.; Zhang, J. A concept of integrated environmental approach for building upgrades and new construction: Pt 1 – setting the stage. *J. Build Phys.* 2015 2015, *38*(4), 360–385.

Bomberg, M.; Kisilewicz, T.; Nowak, K. Is there an optimum range of airtightness for a building? *J. Build. Phys.* 2016, 39, 395–420.

Bomberg, M.; Wojcik, R.; Piotrowski, Z. A concept of integrated environmental approach, part 2: Integrated approach to rehabilitation. *J. Build. Phys.* 2016a, *39*, 482–502.

Bomberg, M.; Yarbrough, D.; Furtak, M. Buildings with environmental quality management, part 1: Designing multi-functional construction materials. *J. Build. Phys.* 2017, *41*, 193–208.

Bomberg, M.; Romanska-Zapala, A.; Yarbrough, D.W. History of American Building Science: steps leading to scientific revolution. J of Energies. 2020 1027, *13*, 5. p.

Bomberg, M.;. Anna Romanska-Zapala, and David Yarbrough Towards a new paradigm for building science (building physics. *World.* 2021, *2*(2), 194–215. https://doi.org/10.3390/world2020013.

Brennan, T.; Henderson, H.; Stack, K.; Bomberg, M. Quality Assurance and Commissioning Process in High Environmental Performance (HEP) Demonstration House in NY State. 2008. online: www.thebestcon ference.org/best1 (accessed on 12 October 2019).

*Information Notes 2016* Rosemount Atelier in Montreal. Canadian Mortgage and Housing Corporation. Ottawa, Ontario, Canada.

Building America, https://www.energy.gov/eere/buildings/articles/building-americaprogram

Buratti, C.; Vergoni, M.; Palladino, D., Thermal comfort evaluation within non-residential environments: development of Artificial Neural Network by using the adaptive approach data, 6th Int. Building Physics Conf., IBPC 2015, Energy Procedia, 2015, 78, p. 2875 –2880

Chitwood, R.; Harriman, L., 2011, Measured home performance, a guide to best practices for home energy retrofits in Califirnia, Version 1.2, Gas technologyInstitute

Covey, S.R.;., The 7 Habits of Highly Effective People, Simon, and Shuster,1989, pp. 95–182.

IDP, 1991, An Integrated Approach to Design of Protocol Specifications Using Protocol Validation and Synthesis, *IEEE Transactions on Computers*, Apr. 1991, pp. 459–467, 40, DOI 10.1109/12.88465

De Deer, R.; Zhang, F. Dynamic environnent, adaptive comfort, and cognitive performance. In Proceedings of the 7th International Building Physics Conference, IBPC2018, Syracuse, NY, USA, 23–26 September 2018; pp. 1–

Dudek, P.; Górny, M.; Czarniecka, L.; Romanska-Zapała, A. IT system for supporting the decision-making process in integrated control systems for energy efficient buildings. In Proceedings of the 5th Anniversary of World Multidisciplinary Civil Engineering-Architecture-Urban Planning Symposium – WMCAUS 2020, Prague, Czech Republic, 31 August 2020.

Dudzik, M.; Romanska-Zapala, A.; Bomberg, M. A neural network for monitoring and characterization of buildings with Environmental Quality Management, Part 1: Verification under steady state conditions. Energies. 2020, 13, 3469.

Dudzik, M.; Toward characterization of indoor environment in smart buildings; Part 1: Using the Predicted Mean Vote criterion. *Sustainability.* 2020, 12, 6749.

Fadiejev, J.; Simonson, R.; Kurnitski, J.; Bomberg, M. Thermal mass, and energy recovery utilization for peak load reduction. *Energy Proceedia.* 2017, 132, 38.

Fedorczak-Cisak, M.; Bomberg, M.; Yarbrough, D.W.; Lingo, L.E.; Romanska-Zapala, A. Position Paper Introducing a Sustainable, Universal Approach to Retrofitting Resididential Buldings. *Buildings.* 2022, *12*(6), 46. https://doi.org/10.3390/buildings12060846.

Ferreira, P.; Silva, S.; Ruano, A.; Negrier, A.; Conceição, E., 2012, Neura Network PMV Estimation for Model-Based Predictive Control of HVAC Systems, WCCI 2012 IEEE World Congress on Comp. Intelligence, June 10–15, 2012, Brisbane, Australia 15–22, DOI: 10.1109/IJCNN.2012.6252365666

Flaga-Maryanczyk, A.; Schnotale, J.; Radon, J.; Was, K. Experimental measurements and CFD simulation of a ground source heat exchanger operating at a cold climate for a passive house ventilation system. *Energy and Buildings*. 2014, *68*(A), 562.

Fort, J.; Kocí, J.; Pokorný,; Podolka, L.; Kraus, M.; Cerný, R. Characterization of responsive plasters for passive moisture and temperature control. *Appl. Sci.* 2020, *10*, 9116. http://dx.doi.org/10.3390/app1024911.

Häupl, P.; Grunewald, J.; Fechner, H. Moisture behavior of a "Gründerzeit" -house by means of a capillary active interior insulation. In Proceedings of the Building Physics in the Nordic Countries, Gothenburg, Sweden, 24–26 August 1999; pp. 225–232

Heibati, R.S.; Maref, W.; Saber, H.H. Assessing the Energy and Indoor Air Quality Performance for a Three-Story Building Using an Integrated Model, Part 1: The Need for Integration. *Energies*. 2019, *12*(24), 4775.

Heibati, R.S.; Maref, W.; Saber, H.H. Assessing the energy, indoor air quality, and moisture performance for a three-story building using an integrated model, part three: Development of integrated model and applications. *Energies*. 2021, *14*(18), 5648. https://doi.org/10.3390/en14185648.

Hu, X.; Shi, X.; Bomberg, M. Radiant heating/cooling on interior walls for thermal upgrade of existing residential buildings in China. In Proceedings of the In-Build Conference, Cracow TU, Cracow, Poland, 17 July 2013, p 314.

Kisilewiicz, T.; Fedorczak-Cisak, M.; Barkanyi, T. Active thermal insulation as an element limiting heat loss through external walls. *Energy Build*. 2019, *205*.

Kisilewicz, T.; Fedorczak-Cisak, M.; Sadowska, B.; Ickiewicz, I.; Barkanyi, T.; Bomberg, M.; Gobcewicz, E., On the results of long-term winter testing of active thermal insulation, Oct 2023, 113412

Lingo, L. Jr.; Roy, U. Novel Use of Geo solar Exergy and Storage Technology in Existing Housing Applications: Conceptual Study. *J. Energy Eng.* 2016, *143*, 0401602.

Kuhn, T.S.; The Structure of Scientific Revolution. IL, USA: The Chicago U. Press, 1970 see also The Fourth Industrial Revolution | Essay by K. Schwab | Britannica, acc. Dec 17, 2023.

Kosuke, S.; Kataoka, E.; Horikawa, S. Thermo-Active Building System Creates Comfort, Energy Efficiency. *J. ASHRAE*. March. 2020, *62*(3), 42–50. ASHRAE.org.

Klingenberg, K.M.K.; James, M. Homes for a Changing Climate: Passive Houses in the U. S. Paperback – January 1, 2009, order through the internet

Klõšeiko, P.; Kalamees, T. Case Study: In-situ Testing and Model Calibration of Interior Insulation Solution for an Office Building in Cold Climate. CESB (Central Europe Symp. On Building). 2016 2016.

Klõšeiko, P.; Kalamees, T. 2018, Long term measurements and HAM modelling of an interior insulation solution for office building in cold climate. 7th International Building Physics Conference (IBPC2018); Syracuse, NY, USA; Sept. 23–23, 2018, Syracuse, U, 1423–1428.

Kosuke, S.; Kataoka, E.; Horikawa, S. Thermo-Active Building System Creates Comfort, Energy Efficiency. *J. ASHRAE*. March. 2020, *62*(3), 42–50. ASHRAE.org.

Lingo, L. Jr.; Roy, U. Novel Use of Geo solar Exergy and Storage Technology in Existing Housing Applications: Conceptual Study. *J. Energy Eng.* 2016, *143*, 0401602.

Mattock, C.; Harmony house Equilibrium project. In Proceedings of the Canada Green Building Council, Annual Conference Vancouver, Vancouver, BC, Canada, 8–10 June 2010.

Meadows, D. The limits of growth, The Donella Meadows project, Academy of System Change, on the internet; see also "Meadows on social paradigms Meadows, D. H.; Meadows, D.L.; Randers, J.; Behrens III, W.W. The Limits to Growth; Universe Books: NY, NY, USA, 1972

Private communications 2018, with IBACOS and BSC teams of BA.

R2000 standards, https://natural-resources.canada.ca/energy-efficiency/homes/professional-opportunities/r-2000-standard-for-builders/20564.

Regina Conservation House 1978, Pamphlet from the Government of Saskatchewan

Romańska-Zappala, A.; Furtak, M.; Fedorczak-Cisak, M.; Dechnik, M. The Need for Automatic Bypass Control to Improve the Energy Efficiency of a Building Through the Cooperation of a Horizontal Ground Heat Exchanger with a Ventilation Unit During Transitional Seasons: A Case Study. *WMCAUS. Prague IOP Conf. Ser, Materials Sci. and Eng*. 2018, *246*.

Romanska-Zapala, A.; Bomberg, M.; Fedorczak-Cisak, M.; Furtak, M.; Yarbrough, D.; Dechnik, M. Buildings with Environmental Quality Management (EQM), part 2: Integration of hydronic heating/cooling with thermal mass. *J. Build. Phys*. 2018, *41*, 397–417.

Romanska-Zapala, A.; Bomberg, M.; Yarbrough, D. Buildings with Environmental Quality Management (EQM), part 4: A path to the future NZEB. *J. Build. Phys*. 2018, *43*, 3–21.

Romanska-Zapala, A.; Bomberg, M. Can artificial neuron networks be used for control of HVAC in environmental quality management systems? In Proceedings of the Central European Symposium of Building Physics, Prague, Czech Republic, 23–26 September 2019.

Simonson, C.J.; Salonvaara, M.; Ojanen, T. Heat and mass transfer between indoor air and a permeable and hygroscopic building envelope: P. II, Verification, and numerical studies. *J. Bldg. Phys*. 2004 2004, *28*, 161–185.

Thorsell, T.; Bomberg, M. Integrated methodology for evaluation of energy performance of the building enclosures. P3: Uncertainty in thermal measurements. *J. Build. Phys*. 2011, *35*, 83–96.

Torrie, R.; Bak, C., 2022, Building Back Better with a green renovation wave, (Planning for a green recovery), internet newsletter, April 22, 2022 (own archives)

Kuhn, T.S.; The Structure of Scientific Revolution. IL, USA: The Chicago U. Press, 1970 see also The Fourth Industrial Revolution | Essay by Klaus Schwab | Britannica, accessed Dec17, 2023.

Vereecken, E.; Roels, S. Capillary active interior insulation: do the advantages really offset potential disadvantages? *Mater Struct*. 2015 2015, 48(9), 3009–3021.

Walburger, A.; Brennan, T.; Bomberg, M.; Henderson, H. Energy Prediction and Monitoring in a High-Performance Syracuse House. 2010. Available online: http://thebestconference.org/BEST2 (accessed on 2 October 2019).

Wright, G.; Klingenberg, K. *Climate-Specific Passive Building Standards*; U. S. Department of Energy, Building America, Office of Energy Efficiency and Renewable Energy, Report, 2015.

Yarbrough, W.; Bomberg, M.; Romanska-Zapala, A. Buildings with Environmental Quality Managment (EQM), part 3: From log houses to zero-energy buildings. *J. Build. Phys*. 2018, 43. 296, Oct 2023, 113412.

Yarbrough, D.; Bomberg, M.; Romanska-Zapala, A. On the next generation of low energy buildings. *Adv. Build. Energy Res*. 2019. /10.1080/ 17512549.2019.1692070.

Yarbrough1David, W.; Bomberg2, M.; Romanska-Zapala, A. On the next generation of low energy buildings. *Adv. Build. Energy Res*. 2021 2021, *15*, A Paradigm Shift in Integrated Building Design - Towards Dynamically Operated Buildings.

# Chapter 1
# Objective and executive summary of the book

The review of passive house applications and trends for retrofitting technology has led to a mix of climate-modified applications, selected with a view to making universal technology.

**Objectives**: Observe that the current generation of building physics is different from that of the twentieth century. It is because of the integrated design protocol (IDP) creating an individual vision for the designed building and the presence of the energy modeler. Our generation already deals with "building as a system; it is now time to expand the system thinking to the new dimensions. The following book describes a new approach called passive and thermo-active cluster system technology, so the objective of this book is to develop and discuss *a set of methods to make new or retrofitted, sustainable construction with zero carbon emission and near-zero energy use,* and we also propose that the objective of building physics is *to analyze and compare different methods used with the same objectives.*

While the book currently identifies the best technology, we need to create a social wave to make this technology popularly used to increase its affordability. To this end, the book proposes establishing a socio-economic consortium, similar to those built in North America in the 1980s and 1990s, but aiming at linking retrofitting technology to slow the rate of climate change.

**Background:** Since BEST1 (Bomberg and Onysko, 2008) conference, a small, virtual team of engineers has been addressing the challenge of integrating heat, air and water transfers with *monitoring and modeling to evaluate and improve the performance of new technology for retrofitted dwellings and buildings.* The challenge has been to extrapolate from today's technology the critical trends for the future, simultaneously developingnew test methods for integrated buildings. The technology presented in this book *combines passive house approach with a new energy supply method,* to achieve a 70% reduction for retrofitting existing buildings, and a 90% reduction for new construction, all within a reasonable cost. The references are based on the residential 2004 standards of a country.

Still, the best technology alone may not impact climate change. The Canadian demonstration of a passive house in 1978 failed because the gap between advanced building science and construction practice was too large. Canada and the USA responded by instituting public-private, high-impact national programs that changed the thinking paradigm from improving materials to designing the whole building and selecting materials for specific contributions to the assembly.

https://doi.org/10.1515/9783112217023-001

## 1.1 Defining the knowledge domain for energy and indoor environment

Prof. Hutcheon (1998) defined building science as a complex needed to achieve predictability of performance and stated that performance-limiting condition or a failure must also be known. The distance to failure becomes our measure of performance:

> Knowledge about building, called, for convenience, building science, is valuable largely because it is useful in predicting the outcome of the result of some building situation . . . Only with knowledge is it possible to assess the relevance of experience and thus to draw upon broader and more varied experience to develop predictability.

Thus, the term "building science" became a substitute for "building engineering" because engineering merges theory and practice. In the US, building science is practiced by engineers and architects acting as industrial consultants and is seldom taught at universities because its multidisciplinary character requires a broad, multidisciplinary background.

Often, courses of "building physics" include a simplified introduction to environmental control, acoustics, and illumination. Such a course lacks relation to building practice and the economics of construction and may contribute to increase the gap between energy efficiency as taught and as achieved in practice.

## 1.2 A gap between energy efficiency intended and achieved in practice

The American Institute of Architects proclaimed a carbon-neutral future, yet practice shows that the savings on space energy are used for occupants' comfort and do not reduce total energy use. As shown in Table 1.1, when space heating energy was reduced during the period from 1978 to 2005, comfort components were increased (Bomberg and Onysko, 2008). Builders used the savings to introduce measures to improve comfort, and total energy consumption remained the same.

**Table 1.1:** Total energy use (10.6 quadrillion BTU) has not changed; space heating was reduced, but comfort components were increased (Bomberg and Onysko, 2008).

| Category | % in 1978 | % in 2005 | Difference |
|---|---|---|---|
| Space heating | 66 | 41 | −25% |
| Water heating | 14 | 20 | +6% |
| Air conditioning | 3 | 8 | +5% |
| Appliance/electronic | 17 | 31 | +14% |
| Total use of energy | 100 | 99.7 | −0.3% |

## 1.3 Failure of the first passive house in the world

The 1978 passive house had technology like the current one (Energy Conservation House, 1978). Yet, it failed because the gap between advanced science and practice was too large. Builders liked an airtight, well-insulated house, but to reduce the cost they replaced the gas furnace with electric baseboard heating. By eliminating the chimney, they changed the pattern of airflow and modified the indoor environment, causing sick buildings (not enough fresh ventilation air), and wet attics (increased humidity and condensation on top floors and attics). This failure led to a swift reaction of the Canadian Building Code – first as a recommendation, and in 1985, mandatory mechanical ventilation was required in Canada. These issues were further addressed in a high environmental performance N.Y. State project (Bomberg et al., 2009; Brennan et al., 2008), where efficient mechanical ventilation designed as a hybrid solution with high-velocity small-diameter, flexible tubing (Walburger et al., 2010) was used.

The gap between science and practice motivated North America (the USA and Canada) to introduce national public-private programs with a significant effort in public education. Those programs changed the situation in North America, and it is the opinion of the authors that today, their repetition is necessary and should be combined with the introduction of a vision for retrofitting (see later text).

## 1.4 Failure of the traditional retrofitting approach

Figure 1.1 from the Lawrence Berkeley National Laboratory (LBNL, 2008; Bomberg and Onysko, 2008) shows BAU that means building as usual (no changes), and proposes an energy reduction for new commercial buildings to 10% and for all existing buildings to 50% of the level in the early 2000s. Today, 17 years later, new buildings are on track, but retrofitting has been a paramount failure. New buildings are designed as a system, while retrofitting is *traditionally fragmented* and based on return on investment.

The division in Figure 1.1 is purely administrative. The economic value of the building depends on its location, functions, and comfort of living.

## 1.5 The need for retrofitting vision

Reviewing the state of the art in retrofitting, we see a few excellent technologies, but none of them is universal. Passive buildings in Germany are different from those in Texas, but those in Texas are different from those in Illinois. Some of the differences relate to the climate, but most come from traditions. One may also wonder why the industrialization of house construction is not moving faster, given the growing gap

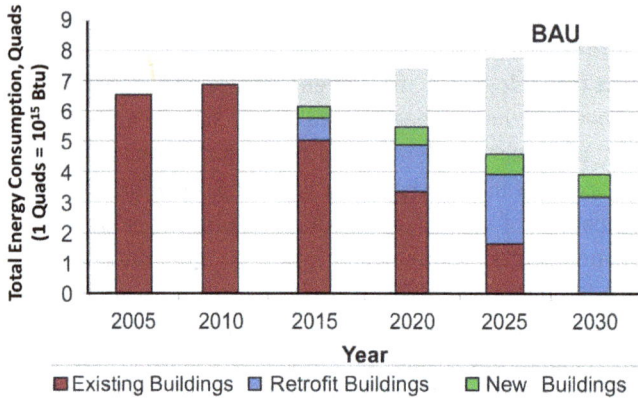

**Figure 1.1:** To stay within the climate warming limit, LBNL forecasted limited new buildings to 10%, and retrofitted buildings to 50% of the 2004 energy codes (Bomberg and Onysko, 2008).

between productivity in the industry versus that in construction. Furthermore, inflation in the post-COVID era has made the cost of housing beyond the reach of the young population.

Looking at the field of automobiles, we notice that a revolution in technology followed the change of thinking paradigm. Henry Ford observed that if the car comes to the worker, the outcome is better. The concept of assembly became universal; mass production reduced the price of the Model T by a factor of 5 in 10 years. The second scientific revolution (Kuhn, 1970), the so-called quality revolution, was based on the Japanese observation that if we do not know why people make mistakes, we should help them avoiding making mistakes rather than studying how many mistakes were made, which is what the Western world was doing since the 1950s.

Thus, we need a vision of retrofitting. Such was also the conclusion of the 2010 Nanjing conference on building physics, because the climate in this region of the world is humid all year round. Control of air humidity alone is easy. Unfortunately, building science in America has been influenced by politics (Bomberg et al., 2020) and in Europe the concept of air recirculation was deleted for the sake of the "green simplification." Now, lessons from pandemics have brought air filtration to the front of indoor environment considerations. Water, with the capacity for heat storage, four times higher than air, must remain as the carrier of thermal energy, but air movement, being involved in most environmental considerations, is the second key component of design for energy and the environment.

Effectively, the vision must include integration of these two subsystems in the process of monitoring and modeling to evaluate and improve the performance of new technology for retrofitted dwellings and buildings.

There is another reason for which the academic community must publish a vision for building renovation, rehabilitation, upgrade, modernization, and many of those words often used by politicians without understanding because, in most cases, their green subsidies in the marketplace are breaking the progress of technology instead of making it easier. As an example, the past administration in Poland was supporting manufacturers of heat pumps and their clients at the same time. What they achieved was a huge increase in the market price of heat hump and the proliferation of worst technical solutions, for example, so-called "heat pump-operated hot water tank" had 81% of direct electrical heater and 19% from the heat pump. If there was a scientifically documented vision for retrofitting, the level of political ignorance would be reduced.

### 1.6 Improving project vision: integrated design protocol

In the 1990s, the uncertainty in design objectives was reduced by a compromise. As predicting energy consumption was beyond the capability of architects and structural engineers, an energy modeling expert was added to the design team. At the same time, the analysis of the environment in buildings was moved to a conceptual stage. This process was called a design charette, as during the French Revolution, charette was the carriage taking the condemned to the guillotine. The IDP concept spread worldwide. Why? Is it because IDP creates a common vision of the building for the design team? Perhaps, but the real reason was economics. By placing the environmental decisions upfront, IDP reduced the cost of design.

## 2 Universal building energy and environmental technology

The oldest piece of research, a confidential report by the National Research Council of Canada (report 1639 from December 1947), provided field measurements during the winter of 1946/47 on two one-story buildings, where metal pipes were placed under the whole ground floor on a partly insulated concrete foundation to provide hydronic heating to the buildings. Each building had a 1.2 m high crawl space above the heated slab and a living space of about 20 m$^2$. The report presented the amount of heat supplied to the buildings as a function of the exterior temperature and the level of ventilation (tests were performed in the ventilation range between 1.5 and 4 ACH).

About 60 years later, a Hungarian inventor, uses a hydronic heat exchanger in an additional concrete slab located some distance under the building. The geothermal hydronic heat exchanger in Hungary did not provide enough energy, and a heat pump was added to the ventilation system below the basement floor, calling it "active thermal insulation" (Kisilewicz et al., 2019, 2023). At the same time, a PhD student at Syracuse University (Lingo and Roy, 2016) who built a heat exchanger circulating air in walls from a split-level heat pump operating on a geothermal water tank in the ground, switches to water-sourced heat exchangers.

About the same time, a team of motivated people in the city of Montreal, Canada, realizing that success requires addressing many design details, formulate the process of integration as a multiple-stage construction (Rosemount Atelier in Montreal, 2016). Over 10 years of stepwise construction they reduced the 90% energy use in 2018, as postulated by the Lawrence Berkeley National Laboratory (LBNL, see conference BEST1, 2008; Bomberg and Onysko, 2008).

The need for universal technology that was made clear during a Western-Chinese building science conference in 2010 in Nanjing created a small virtual network (Mattock, 2010; Bomberg et al., 2011, 2015, 2016a), which already had worked on thermal mass effect in wood-frame home (Mattock, 2010), a demonstration home built by a Hungarian inventor in 2008 (Kisilewicz et al., 2019, 2023), a demonstration building by an American PhD in 2008 (Lingo and Roy, 2016), the Montreal district built in multistages from 2008 to 2018 (Rosemount Atelier in Montreal, 2016), research at Cracow Technical University on neural networks (Dudek et al., 2020; Dudzik et al., 2020; Dudzik, 2020) and on pre-conditioning of ventilation air (Romańska-Zapała et al., 2018, Fedorczak-Cisak et al., 2022; Bomberg et al., 2016), and collaborated with the US passive house institute (Klingenberg and James, 2009; Wright and Klingenberg, 2015), Tech. University of Quebec (Heibati et al., 2019, 2021), and Southeast University (Hu et al., 2013). In Tokyo, Japan, an American technology award winning solution in 2020, referred to as the thermo-active system (Hu et al., 2013), uses both hydronic heating/cooling and ventilation air as a means of energy delivery (or removal). On the moisture control side, the development started in 1999 (Häupl et al., 1999) and continued (Simonson et al., 2004; Bomberg and Pazera, 2010; Vereecken and Roels, 2015); and on indoor climate (Fort et al., 2020; De Deer and Zhang, 2018) completed the first stage of our network activity.

The Atelier Rosemount project in Montreal, Canada (2016), by introducing the multistage construction deals with new broke the barrier of affordability. In the construction quarter, where luxury apartments have exposure on sunny and shaded sides to enable cross-ventilation are neighbored by social dwellings with a cost being a fraction of the previous ones but about the same comfort. The financing concept introduced by this project was a few short-term loans for each new stage of construction, and we call it in the book "a two-stage construction process."

Hungarian inventor whose measurements data were analyzed by Kisilewicz et al. (Kisilewicz, Fedorczak-Cisak and Barkanyi, 2019; Kisilewicz et al., 2023) together with work in Syracuse University, NY, USA, introduced the active part of the passive and thermo-active cluster (PTAC), to which the US National Laboratory (Bomberg and Onysko, 2008) provided quantification for the vision of sustainable construction. In effect, the synthesis of work of Atelier Rosemont, Montreal, Canada, the Hungarian inventor, the Tokyo design team (Kosuke, Kataoka and Horikawa, 2020), and the US National Laboratory (LNBL) can be stated as:

the foundation for universal technology of sustainable buildings with near zero energy and near zero carbon-emission has been established in 2020.

Yet, despite these and other publications (Buratti et al., 2015; Ferreira et al., 2012), the multistage or thermo-active (TA) technologies have not been popularly used. *This indicates a poor linkage between the academic community active in energy and indoor environment and construction practice.*

# 3 Universal energy and environmental technology for retrofitting

Investors must follow the minimum requirements of codes and standards. Society, however, needs a much higher investment level, either net zero energy, or at least near zero energy. The two-stage construction process alleviates this conflict. The second stage of the project will be subject to the same financial restrictions as any other retrofitting project. Either a builder or a house owner who applies for a mortgaged faces not must review two critical issues: (a) the value of the existing property versus surrounding properties and (b) an estimated cost of the planned repairs. Having a cost estimate from stage 1 is invaluable. The next section reports on one successful, multistage construction project.

### 3.1 Multistage construction: Atelier Rosemont in Montreal, Canada

The initial construction began in 2008 and was stepwise upgraded until reaching 92% of the cumulative reduction 10 years later. The retrofitting included the following steps:
- High-performance enclosure; common water; solar engineering resulting in 36% reduction.
- Gray water, the passive measures of energy reduction give 42%.
- Heat pump heating – all passive measures give a 60% reduction.
- Domestic hot water with evacuated solar panels, a further 14%.
- Photovoltaic panels reduce the total energy to a total of 92% reduction.

A multistage construction process, applied to "Atelier Rosemount" in Montreal (Rosemount Atelier in Montreal, 2016), eliminated the difference between new construction and retrofitting by applying the retrofitting to the next stage of each new construction. Energy savings, shown above, agree with the experience gained from the different applications of the Building America program (Bomberg et al. 2009).

### 3.2 Integration enables new energy supply methodology

Heat pumps are the preferred choice because of the energy multiplication effect. Yet, there are different types of heat pumps, and efficient use of the new technology requires using a water-sourced heat pump, together with two water tanks to provide thermal storage (heat capacity). A heating coil is inserted in the water tank, delivering hot water to floor and wall heat exchangers. The water tanks are called: (1) domestic hot water (DHW) tank and (2) cold water tank (CWT). The latter tank functions as a lower terminal of the heat pump.

We distinguish between two types of thermal storage: short-term thermal storage with 14–18-h thermal capacity to equalize daily loads (Fadiejev et al.; Bomberg, 2017), and long-term storage, with weekly equalization of the extreme thermal loads (Bomberg, 2021). Water-sourced heat pumps (WSHP) typically have a performance coefficient higher than air-sourced heat pumps. Yet, switching to a water-sourced system brings a new consideration about the lower temperature limit. For instance, typically when the temperature in the cold water tank falls below 10 °C, an electric heater starts adding energy to the water. Yet, electric heating has an apparent coefficient of performance (COP) of one, while WSHP may have it higher than 4. It makes sense to use hot water instead of electrical heating to add thermal energy to the cold-water tank. We use a 40 L auxiliary water tank (AWT) and a 120 L gray water tank (GWT) capacity (minimum expected for two adults per day for sanitary and hot water use) and the following routine.

Either when the temperature in the CWT falls below 10 °C or when someone starts showering, 40 L from the CWT is transferred to the AWT and later to the GWT. The gray water tank serves for the recycling of water collected from the CWT when the temperature falls below or exceeds the benchmark in cooling operation. The same operation is performed in cold climates when adding to the CWT warm water from a bath or shower. Thus, cold recycled water comes to the gray water tank in winter and hot water will be used for spooling the closet in summer when WSHP is used for cooling of the dwelling. It may look like a paradox, but as adding a gray water tank lowers the volume of water removal and at the same assist in optimization of the value of COP, it makes the system operation less expensive.

### 3.3 The wall heat exchangers: panels or built in situ

With a large area of wall heat exchangers, we will use a low temperature of the heating system for domestic, namely 55 °C. The wall system is about 20–30% more economical than inside the floor. Furthermore, using adaptable indoor temperature enhances the mass effect (Fadejev et al.; Bomberg, 2021). These values were calculated using Energy Plus computer software with typical film coefficients for horizontal and vertical orientations. Furthermore, Hu (Covey, 1989) found that to achieve 90% efficiency in one-directional heating, one should place the PEX tubing on thermal insulation with a thermal resistance of no less than 1 $(m^2 \cdot K)/W$.

There are two methods of heat exchanger construction. In an in-situ application, the PEX tubing is continuous. In a panel application, the PEX tubing connects with snap-on joints. Panels can be made with different materials; for instance, in China, panels were produced with MgO cement and milled (fiberized) bush of rice and wood waste to provide both moisture buffering and elasticity.

**Choice of the heat pump type for the heating unit:** Heat pump selection continues the local tradition. Both in America, air-sourced heat pumps are replacing the old window air-conditioning units, and in Europe, advanced convective-radiative heat exchangers use the use the exterior-placed heat pump. Nevertheless, a continually operated ventilator and a COP typically below 3 are drawbacks of this solution.

Conversely, WSHP with a COP higher than 4.0 provides a few benefits, namely: simultaneous access to hot and cold water throughout the whole year, easy integration of water recycling, integration with hybrid solar panels and ventilation systems, capability for geometrical expansion of the energy system, multiple stages of heating and cooling, and, finally, integration with a new, climatic district network. The drawback is typically the noisy compressor of the geothermal heat pump, and this issue still must be resolved.

## 3.4  Other improvements included in the PTAC technology

Steven Covey (1989), in "7 Habits of the Most Efficient People," highlighted that any process should start with the definition of the required outcome. To achieve a balance between the often contradictory requirements of energy efficiency, occupant comfort, and high quality of the indoor environment, both analog and analytical thinking must be applied in the IDP (IDP, 1991). IDP created a common vision for design based on the whole-building approach, reduced the cost of the design itself, and shifted the design paradigm to building as a system. The design of the whole building, with primary requirements for components and assemblies, implied material selection at a later stage, empowering the polymeric industry to create new multifunctional (Bomberg et al., 2017) materials.

The urban complex in the cold climate of Montreal, where a high level of thermal insulation and air tightness reduced energy consumption by about 40%. An air-sourced heat pump was added to the passive measures in New York State, and energy reduction reached 55% (see construction process (Walburger et al., 2010) and quality assurance (Brennan et al., 2008)). A Ph.D. study that used co-simulation as an improvement for energy modeling in cold climates showed the limit of passive measures with a heat pump application at 60% reduction (Heibati et al., 2021). To reach more than 60%, one needs to use additional measures, for example, geo-solar engineering or modification of the energy supply system. The concept started in 2010 (Bomberg, 2010) with the work progress documented (Romanska-Zapala et al., 2018; Yarbrough et al., 2018) is now included in the PTAC system technology.

The process of air filtration needed to control the spread of microbiological pollution requires ventilation with a variable rate. This can be done with air gaps created between existing walls and a new heating/cooling system by expanding the retrofitting system. PTAC is a universal, climate adapted, public domain technology developed to become a logistic umbrella, in which different options, for example, artificial neural networks, can interact with commercial service and comfort systems. We do not propose a specific, detailed technology, but a blueprint for different practical options. It is for a designer to select the water tank capacity, power of the WSHP that operate in the night only, or extend hours of WSHP operation, use one- or two-step for water buffer system, and means to ensure the minimum temperature of the low HP terminal. If the supply system is too expensive, the designer may increase the level of thermal insulation of the additional walls or introduce phase change/reflective surfaces in the heating/cooling panels. One may also increase the area of retrofitting and cover all interior partitions.

The next element of widening the scope of work is not self-evident. One cannot count on geothermal energy because of space and price restrictions. Instead, one may count on the COP of WSHP, particularly when controlling the temperature of the lower terminal of the WSHP. Furthermore, about one-third of the thermal energy in shower water can easily be recovered if we include the gray water into the system. The next change in PTAC technology is from the air-water heat pump to water-water one, but using two water tanks and independent water pump to separate water heating/cooling from the space air conditioning. The main reason was an increase of COP from 2.5 to 3 for traditional HP to more than 4 for WSHP. Other benefits of WSHP are: presence of hot and cold water throughout the whole year, easy integration of hybrid solar panels, gray water, and district climatic networks, as well as an increase in heating or cooling efficiency.

The backbone of the PTAC technology is the Modeling and Performance Evaluation MAPE) where field modeling provides calibration for numeric or neural network models that, in turn, can be used for building automatics control system. Nevertheless, technology alone, without the specific social demands, will not solve the problem. As we showed above, the best technologies were not accepted by the market. The builder will do what he/she is paid to do. We know that a scientific revolution is now needed for society to gain control on the rate of climate, for economy to create local jobs, and for the occupant to improve the quality of the indoor environment.

## 3.5 Reducing the cost of decarbonization

Torrie and Bak (2022) writing about Canada, highlighted that despite ten million single-family homes and five million apartments with a floor area of 2.1 billion square meters and 65 million tons of carbon emissions per year, where about two-thirds use natural gas, Canada has no pathway to a low-carbon future. At that time each vehicle was produced individually, much as today we do with housing. In today's dollars, the price of a Ford was US$25,000, and the company sold 19,000 cars in this year. Ten years later, with manufacturing experience, the cost of the Model T went down to US$5,000, and Ford sold 941,000 vehicles.

In 2019, Canadians spent more than $60 billion yearly renovating their homes and $30 billion on space heating. A typical deep retrofit with heat pump conversion for a single-family house is at least $40,000. Calculating a 10-year transition to a low energy level with a yearly average of $36.7 billion per year, that is, a carbon emission cost of $141 per ton (about US$100.00). If one assumes that a new technology reduces the price by 30%, with a 30% increase in carbon efficiency, and the volume of retrofitting grows threefold, the cost of one ton of emissions goes down below $10 per ton of emissions.

Yet, more recently the automotive industry went through the next revolution, (a quality revolution). Developed in Japan, the quality assurance (QA) system, with continuous improvement of quality, is the unifying objective of the manufacturing organization. As today we are in the middle of the fourth industrial revolution, we should re-think retrofitting technology.

The above discussion defines our goals. We need to increase the energy efficiency by one-third while keeping the same cost and increase the volume of retrofitting to reduce the cost of 1 ton $CO_2$ by a factor of 10. This is an objective to slow the rate of climate change by retrofitting existing buildings.

## 3.6 Invention of a climatic district system to replace the district heating

This EQM technology introduces an innovation in the climatic district network (CDN) that may also be applied to historic buildings by pairing them with an adjacent standard building. In this manner, the pair of buildings are included in a local district heating/cooling system. District heating, cooling, and ventilation systems eliminate the difference between a single building and a district of the city. As EQM technology includes thermal storage, and water tanks may be located underground, the district climatic system may either be a part of the building or the energy distributing system. The summary of the research on air-earth heat exchangers (Romanska-Zapala et al., 2018) implies 1 m depth in Central Europe. If a low-density (about 10 kg/m$^3$), polyurethane foam fills this line, the foam will be a dry insulation in winter and wet in summer, heat conductor in summer (gradient inwards) and, as such, it will dissipate heat better than dry.

In summary, sending return water, together with preconditioned air, to the next building may reduce operational costs by increasing COP over that of the heat pump alone. Furthermore, as the construction of settlements is less expensive than that of separate buildings, adding a climatic district system increases investment efficiency and would diminish the difference between separate buildings and city districts.

## 4 The benefits of the proposed approach

The benefits of a holistic approach are:
- There are two methods of system construction:
  (1) in situ, with continuous construction of hydronic loops, and
  (2) a panelized system with heating and cooling, with or without ventilation.
- A typical control system needs real-time energy modeling, while our energy models are parametric, that is, they allow comparing the significance of different parameters. To use them for real-time calculation, one must calibrate the model on monitored field data, use co-simulation of individually verified numerical models, or use combined CFD and numerical models.
- Monitoring and modeling for evaluation (MAPE) allows models to be calibrated and subsequently use them for optimization of mechanical devices. Particularly successful are simple and precise artificial neural network models.
- Most of the mid-rises and high-rises use the corridor air pressure correction for the stack effect, which is different on each floor and varies with the seasons. Those differences have a significant effect on indoor air quality. Adding air pressure differences makes a small difference in monitoring costs but a large difference in analysis capability, particularly when a variable rate of air exchange is included in the building design.
- Modeling should be based on a local weather forecast, subsequently adjusting the predicted weather with the information received from monitoring. In effect, modeling of the indoor environment and energy in a high-rise becomes quite complicated.
- This project also introduces a new type of district climatic network (Bomberg, 2021). Return water from building one is used in building two for the lower terminal of the heat pump. This system can be used for two buildings (historic buildings) or the whole region of buildings.
- District climatic network addresses the preconditioning of air and reduces the difference between designing one building or the whole settlement, and thereby opens a new trade-off capability.
- Finally, the complexity of subsystem interactions in mid- and high-rise buildings requires an experimental verification and analysis of all components of the system in all four seasons of the year.

## 5 Closure

In this book, we extend the passive (house) methodology with the use of:

The two-stage (multistage) construction process that modifies patterns of financing:

– Building automatics to control contribution of thermal mass and additional water thermal storage, linked to water-sourced heat pumps, and for new construction, also with solar panels.
– Adaptable indoor climate achieved with HVAC integrated with the building structure and the MAPE system to optimize energy and the indoor environment during the operation of the building.
– CDN to connect the PTAC to the next building in series, between 2 and 200 buildings, for a historic or city district upgrading.

The word "cluster" in the above definition means both a cluster of passive and thermo-active methods, and a cluster of buildings or, in the extreme case, even one building with surrounding ground, where large water tanks could be stored, or may not be stored if the city is already built.

Technical revolution, like Model T of Ford, is needed to retrofit existing buildings. This revolution will provide a win-win solution for society, economy, and the building's occupants. Society wins with slowing climate change, the economy with plenty of local jobs, and occupants with affordable, excellent indoor environments. As builders do only what society wants them to do, the society should demand buildings to have zero carbon emissions and occupants to have a higher comfort of living

Productivity and well-being of the occupants are the main criteria while energy is not easy to observe. While we stress one aspect of many, we give the impression of unbalanced technology. Meadows (1972) said that grants, tax reductions, and sponsorships do not have an impact on a sustainable built environment, while the highest social value, like climate change, has a large impact. Why are we not using it?

The answer is simple; the global market is nobody's market. The society's values are broken in quest for global monetary win and today we must use different ways of rebuilding local socio-economic values. Observe that in the building physics group of a leading Montreal University no one knew about the Rosemount project. Without the involvement of the academic group, the society did not know about the progress, no one has championed the benefits of this project. One may talk about the poor qualifications of these people who want to introduce ecological construction without collaboration with market needs. What is needed is not a money giveaway but a serious work to educate builders by workshops, presentations and professional organizations. Years ago, we had all these tools but because of the lack of systematic technology transfer and having politicians pushing academic projects without the follow up on the marketplace, we lost contact between the academic and construction worlds.

The epidemics deleted the old order, and while the sponsoring products on the market do not bring any progress, the sponsoring organization must start market-like thinking to create market demand. We repeat: *the technology is just one leg, to walk, a human needs two.* The immediate solution should be to use the sponsoring money to build the missing link to allow society to use technology before it becomes old and obsolete.

## References

Bomberg, M.; Onysko, D. Energy Efficiency and Durability of Buildings at the Crossroads, 2008, http://thebestconference.org/BEST1 (accessed on 25 February 2020).

Hutcheon, N.B. The utility of building science, (1971 talk). *J. Build. Phys.* 1998, *22*, 4.

Energy Conservation House. 1978. Regina, Saskatchewan, Provincial Gov, Pamphlet.

Bomberg, M.; Brennan, T.; Henderson, H.; Stack, K.; Walburger, A.; Zhang, J. High Environmental Performance (HEP), residential housing and building technology, for NY state. *A Final Report to NY State Energy Research and Development Agency and National Center of Energy Mgmt.*, manuscript, 2009, USA, 2009.

Brennan, T.; Henderson, H.; Stack, K.; Bomberg, M. Quality Assur. and Commissioning Process in High Environmental Performance (HEP) Demonstration House in NY State. 2008. online: www.thebestconference.org/best1 (accessed on 12 October 2019).

Walburger, A.; Brennan, T.; Bomberg, M.; Henderson, H. Energy Prediction and Monitoring in a High-Performance Syracuse House. 2010. Available online: http://thebestconference.org/BEST2 (accessed on 2 October 2019).

Kuhn, T.S.; The Structure of Scientific Revolution; The Chicago U. Press: IL, USA, 1970, see also The Fourth Industrial Revolution | Essay by Klaus Schwab | Britannica, accessed dec17,2023

Bomberg, M.; Romanska-Zapala, A.; Yarbrough, D.W. History of American Building Science: steps leading to scientific revolution. *J. Energies.* 2020, *13*, 5. 1027.

Kisilewicz, T.; Fedorczak-Cisak, M.; Barkanyi, T. Active thermal insulation as an element limiting heat loss through external walls. *Energy Build.* 2019, *205*.

Kisilewicz, T.; Fedorczak-Cisak, M.; Sadowska, B.; Ickiewicz, I.; Barkanyi, T.; Bomberg, M.; Gobcewicz, E. On the results of long-term winter testing of active thermal insulation. *Energy Build.* Oct 2023, *296*, 113412.

Lingo, L. Jr.; Roy, U. Novel Use of Geo solar Exergy and Storage Technology in Existing Housing Applications: Conceptual Study. *J. Energy Eng.* 2016, 143, 0401602.

Rosemount Atelier in Montreal. *Information Notes;* Canadian Mortgage and Housing Corporation. Ottawa, Ontario, Canada. 2016.

Mattock, C.; Harmony house Equilibrium project. In Proceedings of the Canada Green Building Council, Annual Conference Vancouver, Vancouver, BC, Canada, 8–10 June 2010.

Bomberg, M.;. Anna Romanska-Zapala, and David Yarbrough Towards a new paradigm for building science (bldg physics. *World.* 2021, *2*(2), 194–215. doi: https://doi.org/10.3390/world2020013.

Thorsell, T.; Bomberg, M. Integrated methodology for evaluation of energy performance of the building enclosures. P3: Uncertainty in thermal measurements. *J. Build. Phys.* 2011, *35*, 83–96.

Bomberg, M.; Gibson, M.; Zhang, J. A concept of integrated environmental approach for building upgrades and new construction: Pt 1 – setting the stage. *J. Build. Phy.* 2015 2015, *38*(4), 360–385.

Bomberg, M.; Wojcik, R.; Piotrowski, Z. A concept of integrated environmental approach, part 2: Integrated approach to rehabilitation. *J. Build. Phys.* 2016a, *39*, 482–502.

Dudek, P.; Górny, M.; Czarniecka, L.; Romanska-Zapała, A. IT system for supporting the decision-making process in integrated control systems for energy efficient buildings. In Proceedings of the 5th Anniversary of World Multidisciplinary Civil Engineering-Architecture-Urban Planning Symposium – WMCAUS 2020, Prague, Czech Republic, 31 August 2020.

Dudzik, M.; Romanska-Zapala, A.; Bomberg, M. A neural network for monitoring and characterization of buildings with Environmental Quality Management, Part 1: Verification under steady state conditions. *Energies*. 2020, 13, 3469.

Dudzik, M.;. Toward characterization of indoor environment in smart buildings; Part 1: Using the Predicted Mean Vote criterion. *Sustainability*. 2020, 12 (17), 6749.

Romańska-Zapała, A.; Furtak, M.; Fedorczak-Cisak, M.; Dechnik, M. The Need for Automatic Bypass Control to Improve the Energy Efficiency of a Building Through the Cooperation of a Horizontal Ground Heat Exchanger with a Ventilation Unit During Transitional Seasons: A Case Study", In *WMCAUS 2018*, Prague *IOP Conference Series: Materials Science and Eng.*, vol. 246

Fedorczak-Cisak, M.; Bomberg, M.; Yarbrough, D.W.; Lingo, L.E.; Romanska-Zapala, A. Position Paper Introducing a Sustainable, Universal Approach to Retrofitting Residential Buildings. *Buildings*. 2022, *12*(6), 46. https://doi.org/10.3390/buildings12060846.

Bomberg, M.; Kisilewicz, T.; Nowak, K. Is there an optimum range of airtightness for a building? *J. Build. Phys.* 2016, 39, 395–420.

Klingenberg, K.M.K.; James, M. Homes for a Changing Climate: Passive Houses in the U.S. Paperback – January 1, 2009, order through the internet

Wright, G.; Klingenberg, K. *Climate-Specific Passive Building Standards*. U.S. Department of Energy, Building America, Office of Energy Efficiency and Renewable Energy: Report, 2015.

Heibati, R.S.; Maref, W.; Saber, H.H. Assessing the Energy and Indoor Air Quality Performance for a Three-Story Building Using an Integrated Model, Part 1: The Need for Integration. *Energies*. 2019, *12*(24), 4775.

Heibati, R.S.; Maref, W.; Saber, H.H. Assessing Energy, Indoor Air Quality, and Moisture Performance for a Three-Story Building Using an Integrated Model, Part Three: Development of Integrated Model and Applications. Energies. 2021, *14*(18), 5648.

Hu, X.; Shi, X.; Bomberg, M. Radiant heating/cooling on interior walls for thermal upgrade of existing residential buildings in China. In Proceedings of the In-Build Conference, Cracow TU, Cracow, Poland, 17 July 2013, pp. 314.

Kosuke, S.; Kataoka, E.; Horikawa, S. Thermo-Active Building System Creates Comfort, Energy Efficiency. *J. ASHRAE*. March. 2020, 62(3), 42–50. ASHRAE.org.

Häupl, P.; Grunewald, J.; Fechner, H. Moisture behavior of a "Gründerzeit" -house by means of a capillary active interior insulation. In Proceedings of the Building Physics in the Nordic Countries, Gothenburg, Sweden, 24–26 August 1999; pp. 225–232

Simonson, C.J.; Salonvaara, M.; Ojanen, T. Heat and mass transfer between indoor air and a permeable and hygroscopic building envelope: P. II, Verification, and numerical studies. *J. Bldg. Phys.* 2004 2004, 28, 161–185.

Bomberg, M.; Pazera, M. Methods to check reliability of material characteristics for use of models in real time hygrothermal analysis. In Res. in Building Physics – Proc.1st Central Eur. Symp. Building Physics (eds Gawin and Kisielewicz), Cracow–Lodz, Poland, 13–15 September 2010, 2010, pp. 89–107.

Vereecken, E.; Roels, S. Capillary active interior insulation: do the advantages really offset potential disadvantages? *Mater. Struct.* 2015 2015, *48*(9), 3009–3021.

Fort, J.; Kocí, J.; Pokorný,; Podolka, L.; Kraus, M.; Cerný, R. Characterization of Responsive Plasters for Passive Moisture and Temperature Control. *Appl. Sci.* 2020.

De Deer, R.; Zhang, F. Dynamic environnent, adaptive comfort, and cognitive performance. In Proceedings of the 7th International Building Physics Conference, IBPC2018, Syracuse, NY, USA, 23–26 September 2018; pp. 1–6

Buratti, C.; Vergoni, M.; Palladino, D., Thermal comfort evaluation within non-residential environments: development of Artificial Neural Network by using the adaptive approach data. In 6th Int. Building Physics Conf., IBPC 2015, Energy Procedia, 2015, 78, p. 2875–2880

Ferreira, P.; Silva, S.; Ruano, A.; Negrier, A.; Conceição, E., 2012, Neura Network PMV Estimation for Model-Based Predictive Control of HVAC Systems, In WCCI 2012 IEEE World Congr. on Comp. Intelligence, Brisbane, Australia 15–22.

Private communications with IBACOS and BCI teams of BA, 2018.

Fadiejev, J.; Simonson, R.; Kurnit ski, J.; Bomberg, M. Thermal mass, and energy recovery utilization for peak load reduction. *Energy Procedia*. 2017, *132*, 38.

Bomberg, M. Anna Romanska-Zapala, and David Yarbrough Towards a new paradigm for building science (bldg physics. *World*. 2021, *2*(2), 194–215. https://doi.org/10.3390/world2020013.

Covey, S.R. The 7 Habits of Highly Effective People. Simon & Shuster, 1989, pp. 95–182.

IDP. An Integrated Approach to Design of Protocol Specifications Using Protocol Validation and Synthesis. *IEEE Trans. Comput*, Apr, 1991, 459–467, 40. doi: 10.1109/12.88465.

Bomberg, M.; Yarbrough, D.; Furtak, M. Buildings with environmental quality management (EQM), part 1: Designing multi-functional construction mat. *J. Build. Phys.* 2017, *41*, 193–208.

Bomberg, M.;. A concept of capillary active, dynamic insulation integrated with heating, cooling and ventilation, air conditioning system. *Front. Archit. Civ. Eng. China*. 2010, *4*, 431–437.

Romanska-Zapala, A.; Bomberg, M.; Fedorczak-Cisak, M.; Furtak, M.; Yarbrough, D.; Dechnik, M. Buildings with Environmental Quality Management (EQM), part 2: Integration of hydronic heating/cooling with thermal mass. *J. Build. Phys.* 2018, *41*, 397–417.

Yarbrough, W.; Bomberg, M.; Romanska-Zapala, A. Buildings with Environm. Quality Management (EQM), part 3: From log houses to zero-energy buildings. *J. Build. Phys.* 2018.

Romanska-Zapala, A.; Bomberg, M.; Yarbrough, D. Buildings with Environmental Quality Management (EQM), part 4: A path to the future NZEB. *J. Build. Phys.* 2018, *43*, 3–21.

Torrie, R.; Bak, C. Building Back Better with a green renovation wave, (Planning for a green recovery), Internet Newsletter. April 22 2022, 2022. (own archives).

Meadows, D. The limits of growth, The Donella Meadows project, Academy of System Change, on the internet; see also "Meadows on social paradigms Meadows. In The Limits to Growth. Universe Books: NY, NY, USA, D.H.; Meadows, D.L.; Randers, J.; Behrens III, W.W., 1972.

# Chapter 2
# The history of environmental control in buildings

Building science principles were derived from the observations and performance experience of existing buildings (Bomberg and Onysko, 2002). Since early wood-frame houses were as cold and leaky as their predecessors (log houses), initial improvements in the environmental control of wood-frame construction involved air leakage control.

### 2.1a  Control air infiltration through walls – introduction of building paper

University of Minnesota works on air leakage through frame walls, which led to the acceptance and use of building paper, distinct from roofing products. Placed on the outer side of the wall, the building paper was impeding the entry of air and rain while permitting some vapor to permeate outdoors. It reduced heat losses by limiting air leakage, improved indoor comfort by reducing drafts, and reduced moisture damage to the walls by preventing wind washing in the wall cavities (Bomberg et al., 2016).

### 2.1b  Thermal insulation in the framed cavity

Greig (1922[1]) testing huts demonstrated the value of thermal insulation placed in the frame cavity. Wall cavities became filled with insulation – first with wood chips, sometimes stabilized with lime, and since the 1920s with shredded newsprint, eventually mineral fiber batts. In 1926, pneumatically applied cellulose fiber insulation (CFI) was used to fill the empty cavities of an existing wall. To this end, holes were drilled through plank sheathing. Still, the initial CFI products were not treated with chemicals except for small quantities of lime and boron salts that were added as protection against mold and rot. Nevertheless, except for water stains opposite an external staircase, no water damage was found during the opening of the walls of this house after 50 years of service (Figure 2.1).

The absence of water damage was explained by computer modeling showing the temperature increase caused by the entry of warm indoor air and water vapor condensation. Computer calculations (Figure 2.2) show the amount of condensed vapor initially increases, reaching a peak, and then decreases with a rising air leakage rate.

---

1 Historic reference is an important contribution to building science but may not represent the current building science knowledge; see the list at the end of the book.

https://doi.org/10.1515/9783112217023-002

As the rate of leakage increases, the warming effect increases, reducing the propensity for condensation, and the amount of condensation is reduced. At the extreme level of air leakage, there would be no condensation, though the wall would be energy inefficient. Figure 2.2 shows that water accumulation is small at the very high and low air tightness. The worst is a moderately loose wall. Therefore, the lack of damage shown in Figure 2.1 implies that there was a sufficiently high rate of air exfiltration, enough to avoid prolonged periods of condensation, and the moisture buffer provided by the wood planks and the CFI, supported by the assembly's drying capability, was sufficient to allow periodic condensation.

Still, the presence of thermal insulation in the cavity lowered the temperature on the outer side, increasing the condensation potential during winter.

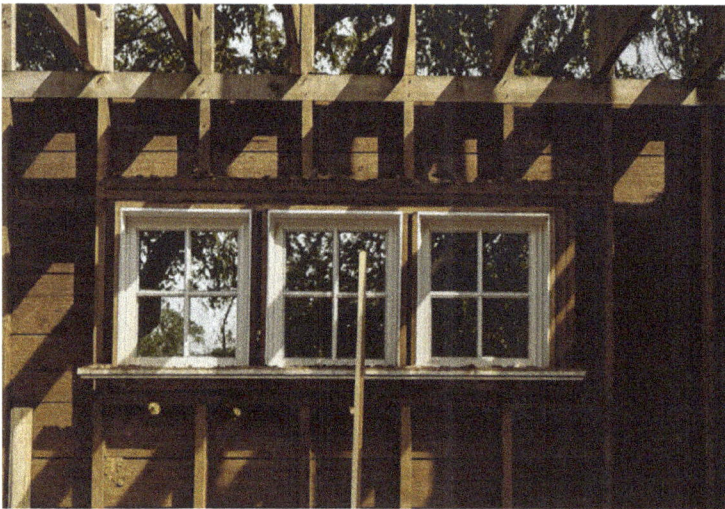

**Figure 2.1:** The wood-frame house was built in 1919 in Saskatchewan, with walls filled with cellulose fiber insulation in 1926, and opened in 1975. No traces of water (Bomberg and Onysko, 2002).

### 2.1c Water vapor retarder on the warm side of the cavity

On the Prairies, already in the 1940s, the walls included cavity insulation, an outside water-resistive barrier, and an interior vapor barrier. The risk of condensation inside wood-frame walls initiated a new area of research. Rowley et al. (1938, HRef) began a study on water vapor movement through insulated walls and, in parallel with Babbitt (1939, HRef), developed a theory of water vapor diffusion through materials. It was established that water vapor diffuses from higher concentrations (partial pressures) to lower, and that condensation occurs when the temperature falls below the dew point. Therefore, a vapor retarder was needed on the warm side of the insulation (Bomberg et al., 2016).

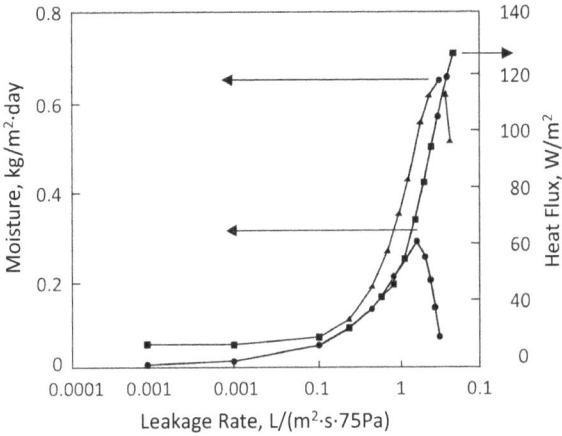

**Figure 2.2:** Heat and moisture in the mineral fiber insulation of the cavity vs the leakage rate of indoor air with 36% and 48% RH (Ojanen & Kumaran, HRef; from Bomberg & Onysko, 2002).

In Canada, vapor barrier has become synonymous with polyethylene sheets, although building codes permit many different solutions. One is a double-painted drywall, in which the paint fulfills the vapor barrier requirement. And the 0.15 mm polyethylene sheathing, which controls air leakage in houses, also functions as a vapor barrier.

## 2.1d  Vapor retarders became mandatory

The quantified vapor condensation gave architects a false impression that water control was addressed. The calculation with the method of Glaser (1958, HRef) was, however, not suitable for decision-making (Bomberg et al., 2020). Condensed water evaporates and continues the diffusion process until reaching a significant change in the resistance to vapor flow. Furthermore, osmotic, capillary, or other forces affecting liquid-phase water are neglected in this approach.

## 2.1e  Introduction of air/vapor barrier

Bomberg and Onysko (2002) noted that 14 papers published before 1960 explained the air infiltration was a primary source of water vapor transport. Wilson and Novak (1959, HRef), analyzing condensation between panes of double windows, showed that with the neutral pressure plane in the middle of the window, air movement carried twice as much water vapor as diffusion, and with the neutral pressure plane at the bottom of the window, this effect was ten times larger than diffusion. Nevertheless, building practitioners neglected this information and ignored all measures to improve air tightness.

The breakthrough came only when practical experience (12 papers in historic references) confirmed the knowledge of experts. What brought about the need for air-tightness was the observation of increased humidity in well-insulated, electrically heated houses. The flue-less houses with high levels of insulation had a much lower frequency of operation of combustion furnaces, creating a concern for indoor air quality. The Canadian Building Code in 1980 required all dwellings to have a mechanical ventilation system, and in 1990, it mandated a minimum ventilation rate of 0.3 air changes per hour (Bomberg et al., 2016a).

The introduction of mechanical ventilation initiated a review of airtightness and the introduction of an air/vapor barrier, which was later separated into air and vapor barriers.

## 2.2 Environmental control: heat, air, and moisture interactions

Nevertheless, heat, air, and moisture (water and vapor) transport across building enclosures are inseparable phenomena. Each influences the others and is influenced by the materials within the building envelope. Often, one simplifies the design process and relates the control of each transfer to a specific material. Thermal insulation is perceived to control heat transfer, and the air barrier to control the air leakage, while the rain screen and the vapor barrier are to eliminate the ingress of water and vapor to materials. Yet, these materials participate in overall system performance and may do more than one function. An air barrier system may provide effective water or vapor control. Thermal insulating material on the outside of a wall (sheathing) may increase the wall temperature and reduce the degree of condensation.

Thus, the environmental control process depends on interactions between heat, air, water, and vapor transfers (Romanska-Zapala et al., 2018), and we must collectively deal with all types of transport. In some ways, this approach represents a return to the old thinking; the difference is the existence of many standards and requirements. Some of them relate to the envelope service and durability over service life, as it must withstand many mechanical and environmental forces.

### 2.2a Moisture effects – material durability

The envelope needs structural integrity and durability to prevent water damage. Of all environmental conditions, moisture poses the biggest threat to integrity and durability, accounting for most damage in building envelopes. Still, many construction materials contain water, most notably masonry or concrete. These materials demonstrate excellent performance characteristics if the humidity does not compromise their structural or physical integrity.

Consider, for example, the ability of a material to withstand, without deterioration, natural periods of freezing and thawing. This so-called freeze-thaw durability is not a material characteristic but a complex property that varies with the material and the environment. For instance, in one school building in South Quebec, only the outer surface of the external clay-brick protrusions showed freeze-thaw spalling. These protrusions were more exposed to driving rain, and the surface temperature of the bricks was slightly lower compared to the plain façade, where no spalling occurred.

Corrosion of metals exposed to air similarly varies with surface temperature and humidity. Likewise, mold growth requires temperatures above 5 °C and relative humidity above 80%.

## 2.2b  Heat energy – dynamic performance

Assessing the energy performance of the building envelope involves three considerations:
- Quantity of heat transferred through the walls, windows, and other elements of the building envelope – the conductive heat transfer
- Quantity of heat needed to bring the temperature of the outdoor air to that of the indoor air – the air leakage characteristics or air exchange rate
- Differences in temperatures on the inner surface of the building envelope – the mold and mildew control points

Conductive heat transfer may be represented in four ways, each with increasing precision. The first approximation considers only the plain, insulated areas of the envelope, ignoring the multidirectional heat flows caused by thermal anomalies. So, a frame wall insulated with Rsi 3.5 glass fiber batt is called Rsi 3.5 wall. The second level of accuracy additionally considers thermal bridges, that is, how the actual thermal resistance of the wall differs from the one-dimensional flow model. Thus, the Rsi 3.5 wall now becomes a Rsi 3.1 wall. The third level of accuracy adds two- or three-dimensional calculations of heat flows, and the fourth level incorporates transient weather conditions into the calculations of the variable thermal performance.

In this way, one differentiates between the declared and design values in the thermal characteristics of building materials. The declared value represented the expected thermal performance measured in a laboratory at a reference temperature and thickness that is stated with a certain level of confidence. The design value estimates the performance under the specified climate and use conditions.

The second component of energy performance – air leakage – relates to the rate of air flowing through the building envelope. There are two different types of air flows, namely ventilation (mechanical, natural, or hybrid) and unplanned air flows (UAF) through the exterior enclosure of buildings. The UAF is directly proportional to air pressure differences across the envelope and inversely proportional to the airflow resistance of the building envelope. The problem, however, is that we do not know the building envelope's transient resistance, nor the air pressure differences, so by using the word "unplanned," we highlight that we make a guess about this component of air change.

The final part of thermal performance evaluation relates to the water vapor condensation on the surface of thermal bridges in the building envelope. At locations with lower thermal resistance, have lower surface temperature in a cold climate winter. As a rule, surface condensation does not plague wood-frame walls, provided the wall cavity is completely insulated and air leakage is insignificant.

Conversely, floor junctions in masonry construction and concrete decks connected to balconies experience condensation problems because of significant surface temperature reductions. If the air movement at these surfaces is low, then mold and mildew in bathrooms and closets, or deterioration of drywall in staircases, can be expected.

Accommodating environmental control in building design requires iterative analysis and a willingness to change concepts until all the consequences of any modification are thoroughly examined. The design of an air barrier system may be an example of an iterative design process. The information flow may start with suitable materials. Typical questions are asked about materials, are their air permeability, ability to be extended, pliability, adhesion, and means of attachment, connection, and support, the long-term performance, material aging, stress, deformations during service, and projected repair and maintenance costs.

After making an initial selection, the designer specifies the architectural details, such as intersections and joints between building elements (foundations, walls, floors, windows, and doors). Then, to achieve satisfactory performance in these locations, the designer must ask further questions concerning the whole system's performance, such as the rate of air leakage, location of leakage, risk of drafts, and impact on condensation. Throughout the design process, the designer consults with structural, electrical, and mechanical experts to obtain answers to all these questions and to ensure that the selected materials will perform satisfactorily.

In addition, the designer reviews the buildability aspects, such as material installation under different weather conditions, the level of labor skill required for installation, and construction tolerance. Buildability, as the word suggests, reflects whether the design on paper can be constructed. Finally, the complexity of heat, air, and moisture interactions demands redundancy in the design. For instance, the air barrier plan may be punctured, not connected to some elements of the envelope, or a rain leak may develop. The designer must evaluate how the moisture could be drained or, if not drained, whether it could be dried out. How long would the drying take, and what effect would it have on other materials? Could the prolonged presence of moisture cause corrosion, mold growth, or rot?

The entire process of environmental control design must occur off-site and never at the building site. Addressing only a specific design problem on the job site without reviewing all the performance effects may cause problems, as other requirements may not be achieved.

## 2.2c  Addressing the duality (plurality)

In designing for environmental control, professionals integrate two very different conceptual processes. One involves specific testing and analysis; the other encompasses broad qualitative assessments based on experience, judgment, and knowledge of what makes a building envelope function. On the analytical side is a complex array of tools, models, and data which describe the material, structural, and environmental factors relating to the building envelope. On the qualitative side is a sense of how a particular building envelope would function.

For example, a vapor barrier is typically classified at one perm, a unit that represents sufficient retardation of water vapor flow for wood-frame housing. However, if calculations were made using a complex model of heat, air, and moisture transport for various climatic conditions, for instance, in Canada, water vapor retarders ranging from 0.1 to 10 perms would be suitable for multiple combinations of materials and climates. Similarly, some materials that qualify as vapor barriers.

The beginning of Building Physics dates back to the 1930s in Russia, when "Stroitielna Tieplofysika" was first published. With a focus on heat and moisture performance in various applications, Russian literature exploded in the 1950s and 1960s. During this time, thanks to the work of J.S. Cammerer in Germany and H. Johansson in Sweden, we got "Baufysik" and "Byggnadsfysik," respectively. Soon, all these trends merged with the Architectural Science from UK and Australia. Thus, in the 1970s, the academic discipline of Building Physics was firmly established in Central and Northern Europe.

In a paper reviewing the history of Canadian residential housing (Bomberg and Onysko, 2002) wrote:

> This paper reviewed the historical background leading to the current holistic approach to heat, air, and moisture control of building envelopes. It showed that past building industry empiricism (learning from the field observations and forensic studies) was a slow process. While there is no schism between building science and building physics, we must be focused on field performance.

In the summary of the 1st Building Enclosure Science and Technology (BEST 2008) conference, we wrote:

> The building industry is at a crossroads, and the question is, "where do we go from here?" The "green" train has left the station, but the tracks are still being built. At the far end, there is an American Institute of Architects' commitment to achieving a carbon-neutral 2030 future. In the beginning, just outside the station, there is much goodwill but also the realization that most of the highly inefficient buildings of today will be with us well beyond 2030. A chasm must be bridged if the goals are to be achieved, and there still is confusion about how we can accelerate the renewal process.

The word confusion in the above passage is incorrect, our problem is the lack of integration of different pieces of research; a good example is the concept of thermal mass.

## 2.3 Energy mirage demonstrates the role of thermal mass

The average energy use of multi-unit residential buildings in Vancouver in 1990 was 315 kWh/m$^2$ and declined, reaching 250 kWh/m$^2$ in 2002. This would be considered all right, if not for the fact that someone found the average energy use in similar buildings in the 1920s was the same. So, the masonry building without any insulation, built 80 years earlier, consumed as much energy as a shiny, glass-clad building constructed in 2002. This, of course, says very little about the increased functions that current modern facilities are fulfilling compared with those in the 1920s.

In 2002, office buildings, office equipment, and computers used 10% of total energy, but energy for lighting was 28%. For a non-specialist who believes that in 1920 people also used lights (and probably less efficiently than now) and who knows that today we have thermal insulation, air barriers, and many other energy-saving measures, it isn't easy to understand why we do not use less energy than was used in the 1920s. Furthermore, someone observed that from the 1920s to the 2000s, there was a threefold increase in emissions attributable to buildings.

This is the energy situation in which we find ourselves, and in which we, as a society, have agreed to return to carbon-neutral construction, last seen on this continent in the mid-nineteenth century. To see how this can be accomplished, we need to review the changes in building construction over the last 60 years.

The construction of a masonry building took a long time. As the load-bearing function required thick masonry walls on the lower floors that became much lighter on the top floors, such a building had a substantial thermal capacity. The floors contained steel beams and masonry blocks. The walls were airtight because of exterior and interior lime-based plasters (stucco). Lime develops strength slowly, allowing for the settlement and movement of the walls. Excellent, heavy, and typically oil-painted wood of double windows was integrated with the masonry walls. The windows were small. From a building physics point of view, the building was airtight, massive, and well-integrated. Because of its inefficient and periodic heating

sources, the indoor temperature varied between periods of comfort and discomfort.

## 2.4 The failure of fragmented approach

There were two trends. One involved material improvement. While positive for the
material, each change may have produced unpredicted effects on the building. For instance, the elimination of wood strapping under exterior plasters and replacing lime
with cement increased cracking and affected moisture management; the introduction
of simplified methods of window mounting dramatically increased air leakage of the
walls, etc. The other trend was coalescent around the improved indoor environment
and occupant comfort. Increased comfort caused an increase in energy use.

### 2.4a  Improvements in heat, air, and moisture control in cold climates

In summary, many significant design changes have occurred over the last 60 years:
1)   Increased levels of thermal insulation
2)   Increased level of water vapor resistance
3)   Increased air tightness of the walls
4)   Reduced moisture buffer capability
5)   Introduction of more moisture-sensitive materials

Each and all of these can dramatically reduce the moisture tolerance of residential
walls.
     Before the 1970s, our society was not concerned with the cost of energy. Excessive
air leakage and associated heat losses were more inconvenient than serious problems.
The introduction of energy conservation to reduce the impact of oil imports first
started in the 1970s. As we introduced high-efficiency heating devices that eliminated
the need for chimneys, we introduced a new concern – the need for air redistribution
within the home. The interaction of the building enclosure with the heating, ventilation, and air redistribution systems in the occupied space has become part of the
builder's design framework, and a phrase such as "building as the system" describes
it best.

### 2.4b  Energy use in the building systems

Figure 2.1 shows data from the U.S. Energy Information Administration on the total residential energy use in 1978 and 2005. The total energy consumption in the building sector

in 2005 was the same as in 1978, while the fraction of space heating was reduced from 66% to 41%. The improvement of comfort components nullified the savings.

This observation was confirmed during the "energy conundrum" event: in 1929 and 2002, large residential buildings in the city of Vancouver used the same amount of total energy. The fact that heavy masonry buildings without insulation could provide a reasonable indoor climate highlighted the role of thermal mass. Research on the adaptable comfort approach showed that one can modify the indoor air temperature by at least 6 °C over 6 h. The change in human comfort does not exceed 4% of the comfort index. It justifies switching to the adaptable indoor comfort approach and introducing two categories of thermal storage: (a) short term (14–16 h) with the help of the building mass to equalize daily thermal loads, and (b) long term (168 h) with a geothermal or water tank storage facility.

The second conclusion that one can draw from these comparisons is that total energy use is the only measure to be used. Measuring the thermal performance of a component under arbitrary conditions, such as steady-state transmittance (U-value) or resistance (R-value), only introduces confusion and may lead to misjudgment. A well-known example is the case of log houses that did not meet the standard requirements while providing an excellent indoor climate. People do not remember that these standards were created in the 1950s for different walls and different service conditions. Since that time, many small changes have modified building performance, and specifically those introduced after the energy crisis in the 1970s have made these old methods obsolete. To design low-energy buildings, one must use only the correct evaluation methods.

While the media talk glibly about using renewable energy, they do not understand that renewable energy must be integrated into an existing system. First, one must reduce the **gross energy inefficiency** of today's existing buildings, then reduce energy consumption while retaining a good indoor environment, and finally integrate renewable sources of energy into the building system.

When buildings were leaky and poorly insulated, the effect of heating, ventilation, and air conditioning (HVAC) systems on air pressure and the durability of the enclosure were insignificant. There was no need to control air movements except for ventilation. That is not the situation today. Now, in well-insulated, air-tight buildings, we design air movement to minimize potential health problems from mold and microbial contamination.

## 2.4c The failure of fragmented approach to retrofitting

In 1976, architects at the University of Illinois defined an airtight, super-insulated house, and Canadians built a demonstration house. The 1978 passive house used technology almost like the current one. It failed because the gap between the frontiers of building science and building practice was too large. Builders did not realize that eliminating the chimney would change the pattern of airflow and, in turn, modify the whole-house indoor environment. Sick buildings (not enough fresh ventilation air) and wet attics (increased humidity and condensation on top floors and attics) were problems introduced by the elimination of chimneys. Yet, this failure led to mandatory mechanical ventilation in Canada.

The second failure of the fragmented approach can be seen in the 2008 forecast for new and retrofitted buildings. The new commercial buildings were supposed to have 90% energy use reduced, and all existing buildings have retrofitted to 50% of the level in the early 2000s (Figure 2.2). The LBNL forecast was based on practical cases, and the question was whether there had been the will to do it. Today, 16 years later, new buildings, designed as a system, are on track, but retrofitting based on return on investment and the *fragmented approach* is a total failure. Nevertheless, the economic value of a building depends on its location, functions, and comfort of living, and the obsolete administrative terms should not be considered.

## 2.5  Evaluation of systems, not materials

It is essential to analyze the performance of assemblies, not materials. Dealing with materials is easier, and building codes/standards ascribe a specific function to a particular material, as this is the only way a prescriptive code can work. We have functions like water vapor retarders, air barriers, thermal barriers (fire protection), and rain-screen, mentally coupled with materials. But some materials may have different functions; for example, closed-cell polyurethane foam can be used as insulation, a rain screen, a water vapor retarder, or an air barrier. An opposite situation arises when all elements in the assembly provide airtightness or thermal resistance.

Both the architectural design and construction processes are holistic, though they involve many highly specialized people – how should they collaborate during this process? Furthermore, the outcome of an architectural design depends on interactions between different trades involved in installing materials in an assembly. This design aspect is so crucial that we stress the importance of mock-up evaluation and continuous commissioning as separate activities in the construction process. This ensures that the design concept is buildable and that all trades learn what they must do to satisfy the building objectives.

## 2.6  Review of the design principles

Bomberg et al., 2016, with permission of Adelson and Rice (HRef), expanded their work and proposed the following design principles:

**Table 2.1:** Design principles for building enclosures.

**Objectives**
A1. Provide continuity of functions
A2. Provide redundancy of design (second line of defense)
A3. Integrate interactive effects
**Constrains**
B1. Consider separate lives of components or assemblies
B2. Consider the flow of fluids and energy from high to low potentials
B3. Consider moisture-originated deterioration mechanisms
**Balance**
C1. Keep balance between continuity and separation
C2. Assess heat, air, and moisture flows and their effects
C3. Use economic considerations for interactive effects

In designing a building, we need to consider both objectives and constrains and try to establish a balance between them. In traditional masonry construction, all functions were achieved by a composite of masonry and plasters. Emergence of framed and layered structures initiated a process, accentuated by codes and standards, to link each material to its main function in the assembly. Such an approach, however, confuses people, making them forget that all systems always perform as an entity.

**Objective A1: continuity of functions (continuity of performance attributes)**
We need to achieve continuity of all environmental functions (sound, heat, air and moisture transfer, fire and smoke protection, etc.). A funnel can help to visualize the effects caused by a deficiency in this principle. In a narrow part of a small funnel, water runs faster than in the wide one, but when we increase the size of the funnel and fill it with a high-water height, water runs amazingly fast. Similarly, a thermal bridge in a well-insulated wall has a much higher impact on the wall's thermal performance than the same thermal bridge in a poorly insulated wall; or a hole with same size has a much higher impact on air tightness if the wall is very tight.

**Objective A2: second line of defense (redundancy)**
Since buildings are erected in uncertain weather conditions with different materials and may encounter different deficiencies during design or workmanship stages, Adelson and Rice (1991) introduced a principle called "creative pessimism" that we re-named to "the second line of defense." This concept follows the recognition that uncertainty is caused by the variability of materials, workmanship, and weather. One, therefore, requires applying two different measures of control. The need for the second line of control is most visible in the moisture management of cladding systems.

There is a saying: "a perfect design exists only on paper," in the real world, sooner or later, something goes wrong. The failures of the "face seal" approach in moisture management of cladding (stucco or EIFS) highlighted the risk of neglecting the possibility of failures. Note that EIFS stands for Exterior Insulation and Finish System, which is a type of cladding used on the exterior of buildings to provide insulation, a finished surface, and weather resistance. Also, EIFS is a multi-layered system that includes insulation board, a base coat, reinforcing mesh, and a finish coat. EIFS is known for its energy efficiency, durability, and design flexibility. As sealants were the only measure to control water entry, the water-resistive barrier was added as the second line of defense.

**Objective A3: integrate interactive effects**
This principle applies when a final effect can be achieved by a different combination of environmental factors, for example, the temperature of the indoor air depends on thermal mass, thermal insulation, air infiltration, air ventilation, the fraction and orientation of windows, and outdoor weather. Changing one of these factors may affect others and modify the final effect. People who calculate the effect of adding thermal insulation by assuming a constant correlation make a mistake. This principle tells us that any change in the interacting factors must be evaluated in the context of the whole building with a view to both technology and economics.

When trying to fulfill these three objectives, we encounter the following three constraints:

**Constrain B1: consider separate lives of components or assemblies**
Materials have different thermal and moisture expansion coefficients, exhibit differential durability, and may also exhibit a lack of chemical compatibility with one another. This may be a problem if moisture is accumulating in their interface. Consider a joint between the exterior plaster and the rough opening of a window. Typically, fresh Portland cement plaster is applied directly to the rough opening frame. Yet, as with all cement-based materials, it will shrink away from the window opening frame, developing a small crack. This crack will draw water inward (from the wall surface), and the water comes in contact with wood that is a moisture-sensitive material. Historically, when plaster was made of lime or lime-cement, it allowed good drying from the surface. Today, plaster often contains hydrophobic (water-repelling) agents to slow both the rate of water entry and the rate of water drying. This type of failure has been frequently seen in warm and humid climates.

So, if you want to use modern acrylic finishing plaster, you must respect the principle of separate lives and place a gasket (or sealant on a backer rod) between the plaster and the rough opening of the window, or do it in a traditional way, having a brick veneer covering part of the window frame.

**Constrain B2: high to low (follow the gradients)**
This principle relates to energy and mass flows: heat, air, water, vapor, or electric current all flow from high to low potential, being it temperature, pressure, or substance concentration. This law also applies to materials that have been enriched during manufacturing; for example, an oriented strand board panel will reverse to wood fibers. In the latter case, one talks about the durability of materials under the effects of environmental factors, where the rate of damage depends on the severity of exposure.

Examples of the high-to-low principle are: the shedding of rainwater flowing under the action of gravity, the need for a drop edge under windows, and the need for the overlap of water-resistive barriers.

**Constrain B3: consider moisture-originated deterioration mechanisms**
Moisture has not been a consideration in the traditional, massive masonry walls that had a large capacity to absorb and store water. This constrain has been added because today most materials, even masonry walls, lack moisture storage capability and must be considered as damage-prone.

**Balance between objectives and constrains**
We need to achieve a balance with changes in the outdoor environment and the near-constant indoor environment; we need to achieve balance between the various materials in the assembly to avoid distortions and deformations, and similarly, a balance between different components of the building. A good example of design with balance in mind is plywood, with oriented strands going in two different directions. Another example is traditional three-coat stucco, where, starting from the substrate, each layer has higher water vapor permeability and lower mechanical stiffness to avoid warping of the stucco under wetting or drying conditions.

**Balance C1: keep balance between continuity and separation**
Often, the continuity of function can be achieved by an adequately designed discontinuity, for example, by using an overlap of roofing tiles or the use of flashing to compensate for the effect of "separate lives." Other examples may include movement joints, for example, when two adjacent panels can expand and contract, but the joint maintains the connection, or reveals in plaster where a thin section will crack, but the crack will not be visible because the shape of the profile hides the crack, and the underlying material (substrate) will provide the continuity of function.

**Balance C2: use risk assessment for flows and their effects**
A good example of risk assessment is the requirement in ASHRAE Standard 160 that assumes that 1% of the rain load has passed the first layer of defense, and one must calculate whether the specified wall system, in the given climatic conditions, has an adequate capability to dry this moisture within one year.

**Balance C3: use economic considerations for interactive effects**
This requirement has been added because one often selects one parameter out of many when dealing with an interactive situation, for example, increasing thermal insulation or air tightness without consideration what effect it has on other factors. This approach often results in a large gap between design intentions and the practice of construction. An example is the requirement of the nominal thermal resistance of the opaque part of the wall in high-rise buildings, Rsi 4, while the average thermal performance, including windows and air infiltration divided by the wall area, makes an effective R-value index about 0.7 (Cianfrone et al., 2016, HRef). Using economic evaluation, we would invest in more detailed design of large window-wall interfaces and improvement of all thermal bridges, and reduce the requirements for the opaque part of the wall. This is a point supporting total energy use and cost rather than arbitrary requirements in building codes. Building codes do not regulate engineering but health and safety of construction; goodwill without understanding is not the basis for any building code.

    Finally, we have established that the future belongs to green buildings. What considerations make the complex called the "green value" of a building?

## 2.7  Key components of "green value" of the building

In 1990s we have re-discovered the significance of the air-water transport interaction that was first established for windows in Germany in the 1930s and later forgotten. Our understanding of wall/window interface performance is since completed. The list of critical features of "green value" during the design and construction of facilities in the 1990s may have included the following considerations:
1.  Design for durability
2.  Design for energy efficiency and the efficient use of materials:
    a.  Separation of ventilation/air distribution and heating/cooling systems
    b.  Solar-water-cooled, integrated domestic hot water systems
    c.  Heating/cooling unit consisting of water-sourced heat pump (WSHP) and two thermal storage tanks, plus water pump to distribute energy to the space heat exchangers

      d.   MAPE (modeling and performance evaluation) system to calibrate hygrother-mal models and optimize HVAC operation under service conditions (Bomberg et al., 2018)

      e.   Increased use of daylighting

      f.   Indoor environment analysis with a view to occupant health and productivity

      g.   Flexibility, that is, lower costs associated with changing space configurations

      h.   Re-use of materials in building enclosure systems

      i.   Design from cradle to cradle, that is, a reuse of materials at the demolition

3.   Design with efficient use of renewable resources:

      a.   Analysis of inter-zonal and interstitial air flows

      b.   Prediction of building enclosure performance under service conditions

      c.   Use of only those renewable resources that are economically justified

      d.   Design sufficient thermal mass connection with the heating/cooling system for 14–16 h of thermal storage needed for daily load equalization

      e.   Design sufficient capacity in the hot water tank or domestic hot water tank to provide 168 h of thermal storage needed for weekly load equalization

      f.   Design optimum or sufficiently efficient conditions of service for water-sourced HP

4.   Testing the mock-up of building enclosures for commercial buildings

5.   Including the commissioning process as a part of the design and construction process:

      a.   Troubleshooting design drawings is the first step in commissioning

      b.   Testing air leakage as a QA during construction

      c.   Testing fresh air delivery in all rooms after a few weeks of occupation

      d.   Testing air quality of an occupied space after a few weeks of occupation

Number 1 in the green value complex is durability. If one can extend the service life of a building, for example, by 20% longer, one reduces the annual costs and saves materials that would be used in the replacement building. In this process, the direct savings on energy and replacement materials provide a multiplier effect comparable to injecting three to five times more money into the local economy.

    The second critical consideration is to apply all passive measures before progressing to renewable energy generation (Bomberg et al., 2020). The passive measures that are often neglected, even though they offer the most value for the money invested, are:

1.   Consider thermal mass placement that respects the climate and heating/cooling system

2.   Increase airtightness between floors in any building to any possible level

3.   Keep reasonable airtightness of the wall (see Bomberg et al., 2016a)

4.   Increase exterior insulation and reduce thermal bridging

5.   Use solar panels and geothermal or water tank thermal storage

6.   Improve the selection of windows

Translating this list into specific subsystems, for instance:
1) Use geothermal air pre-heat for ventilation
2) Use solar support for the domestic hot water system
3) Use hybrid solar panels coupled with a heat pump and long-term heat storage
4) Use the hydronic system for linking sub-systems like water recycling

Specific technical measures follow these examples:
1) Use a WSHP with radiant heating/cooling
2) Use a dedicated outdoor air system for ventilation
3) Use building partitions as heat exchangers
4) Use diagnostics for malfunctioning a system component
5) Use advanced equipment in illumination, motors, and pumps

In short, we stress that high-performance building enclosures are a prerequisite for the next generation of HVAC and lighting systems. Efficient energy management can dramatically reduce thermal loads, encouraging the use of distributed HVAC systems.

One can also observe a trend for building enclosures to become multi-functional (Straube and Burnett, 2005; Bomberg, et al., 2017). Dynamic envelopes can preheat or pre-cool indoor air (Bomberg et al., 2021). Furthermore, using filters and dehumidifiers on the air supply can also modify the indoor environment. Advances in window technology permit the use of increased daylighting. With reduced thermal loads, several technologies previously discarded in research are becoming more economically viable. Those include effects of thermal mass and phase change materials – even though these effects are climate-dependent – they are coming back as significant improvements in the technology mix.

In a nutshell, designing a building as an integrated system produces a better and less expensive structure than a traditional material selection for a preconceived building. In the next part of this monograph, we will examine which of these concepts were further expanded in the next generation of technology.

# References

Bomberg, M.; Onysko, D. Heat, air and moisture control in walls of Canadian houses: a review of the historic basis for current practices. *J.Build. Phy.* 2002, *26*, 3.

Bomberg, M.; Kisilewicz, T.; Nowak, K. Is there an optimum range of airtightness for a bldg? *J. Build. Phys.* 2016, *39*, 395–420.

Bomberg, M.; Kisilewicz, T.; Mattock, C. Methods of building physics. Cracow University Press, see publication on the internet, 2016a, pp. 1–300.

Bomberg, M.; Yarbrough, D.; Furtak, M. Buildings with environmental quality management (EQM), part 1: Designing multi-functional construction materials. *J. Build. Phys.* 2017, *41*, 193–208.

Bomberg, M.; Romanska-Zapala, A.; Yarbrough, D.W. History of American Building Science: steps leading to scientific revolution. *J. Energies.* 2020, *13*, 5. p. 1027.

Bomberg, M.; Romanska-Zapala, A.; Yarbrough, D. Towards a new paradigm for building science (building physics). *World*. 2021, *2*(2), 194–215. https://doi.org/10.3390/world2020013.

Romanska-Zapala, A.; Bomberg, M.; Yarbrough, D. Buildings with Environmental Quality Management (EQM), part 4: A path to the future NZEB. *J. Build. Phys.* 2018, *43*, 3–2.

Straube, J.; Burnett, E. Building Science for Building enclosures. Building Science Press: Westford, MA, 2005.

# Chapter 3
# Air transport

Air transport is involved in almost all situations analyzed in building physics/science.

## 3.1 Air flow through porous materials

The flow of gas takes place in the whole volume of open-cell porosity. As a gas molecule moves from one pore to the adjacent one, it does not follow a straight line. The connectivity of the cells in the structure leads to a so-called "tortuosity" factor, that is, the ratio of the actual length of the gas path to the minimum distance across a rectangular specimen.

Two different sets of engineering equations are commonly used: The first approach uses a mass-permeability coefficient and the pressure gradient to write a rate-of-the gas-flow equation:

The second approach relates the velocity of the air flow and the pressure gradient Pa/m. The air permeability coefficient, $\delta$, involves an intrinsic permeability coefficient $\kappa$ [m$^2$] and the dynamic viscosity of the flowing medium $\eta$ [Pa·s]:

$$\kappa = \delta / \eta \tag{3.1}$$

## 3.2 Examples of air leakage tests

Table 3.1 highlights the effects of nonlinearity in cracks and openings on the example of a volumetric airflow rate, l/s, through round holes with diameters ranging from 10 to 40 mm, and Table 3.2 highlights the effects of different types of masonry (a) aerated autoclaved concrete (AAC) and (b) 25 cm thick clay brick.

**Table 3.1:** Area and flow rate at 5 Pa pressure difference through holes with diameters ranging from 10 to 40 mm.

| Hole dia. | A | Flow |
|---|---|---|
| mm | cm$^2$ | L/s |
| 10 mm | 0.8 | 0.3 |
| 20 mm | 3.1 | 1.7 |
| 30 mm | 7.1 | 2.3 |
| 40 mm | 12.6 | 4.3 |

https://doi.org/10.1515/9783112217023-003

**Table 3.2:** Air tightness of an aerated autoclaved concrete (AAC) and clay brick masonry (25 cm considered as the reference level) alone and with different renderings.

| Wall system | Air pressure | Difference |
|---|---|---|
| Description | 5 Pa | 50 Pa |
| Description | 5 Pa | 50 Pa |
| Clay brick masonry 25 cm | 100% | 550% |
| + Two sides lime plaster | 15% | 150% |
| Clay brick masonry 38 cm | 50% | 400% |
| + Two sides lime plaster | 12% | 120% |
| Aerated autoclaved concrete | 145% | 700% |
| + 3 mm thin polymeric finish | 10% | 60% |
| AAC + adhesive on surfaces | 3% | 25% |
| AAC + adhesive and 3 mm polymer | 1.8% | 13% |

One can observe how important two-sided purging (plastering) on any type of masonry wall is, and that precision-trimmed aerated concrete blocks with adhesive and polymer-modified plaster make AAC masonry much tighter than standard masonry can ever achieve it.

## 3.3 Air pressure distribution in a building

A natural convection means that a temperature difference causes buoyancy originating air movement; while forced convection takes place when a mechanically induced air pressure difference drives the air. One may also talk about natural or mechanical ventilation, the latter case relating to mechanical devices causing the air pressure difference. Nevertheless, air redistribution in the building may be either partly mechanical ventilation, natural, or hybrid phenomenon.

Figure 3.1 shows an idealized picture of the natural air exchange between the building envelope and the external environment under natural convection conditions. This air movement is called a stack effect.

Let us consider an airtight enclosure, except for the window-wall interface. Temperature difference will cause a linear pressure distribution with over-pressure above and under-pressure below the neutral plane. If air leaks below and above the window are identical, the neutral plane will be in the middle. The pressure difference is a function of the internal and external air density ($\rho$) difference, the distance from the neutral axis (h), and gravitational acceleration (g):

$$\Delta P = (\rho_i - \rho_e) \cdot h \cdot g \tag{3.2}$$

Of course, Figure 3.1 is a hypothetical case of a one-story building. In a multistory building, exfiltration takes place on higher floors and infiltration at the lower floors,

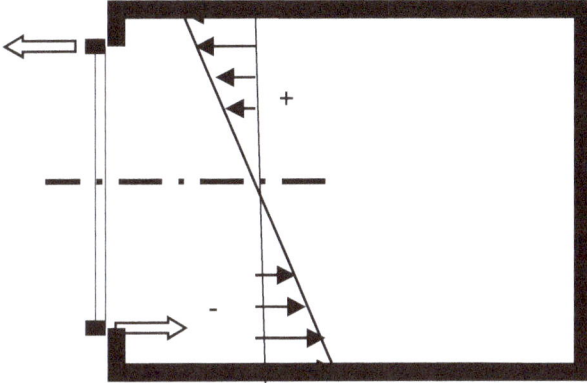

**Figure 3.1:** Air pressure distribution in an airtight room (window without mechanical forces or wind).

as confirmed using infrared imaging. Warm air exfiltration (increased temperature – red or light-yellow colors) can be observed in Figure 3.2.

Pressure distribution in a high-rise building depends on the distance from the neutral axis, and in the case depicted in Figure 3.3, the neutral plane is located at the top of the ventilation chimney.

**Figure 3.2:** Exterior walls with warm air leakage in the window frame. The color scale indicates a probable air pressure difference in Pascals – (Bomberg et al., 2016).

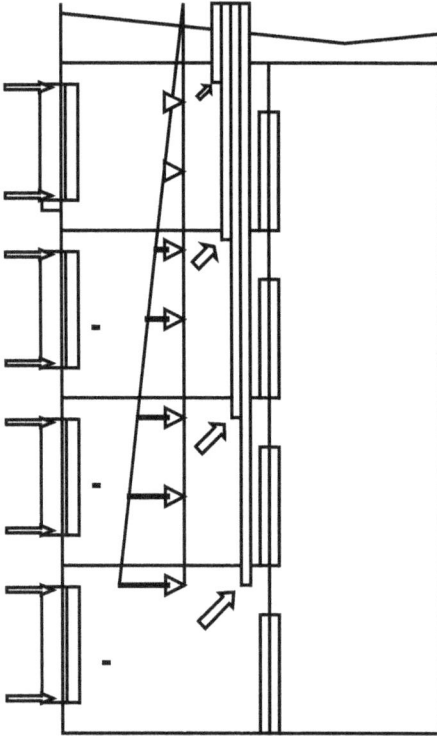

**Figure 3.3:** Scheme of natural ventilation operation in the building.

Stale air is exhausted by the ventilation shaft, and fresh air is supplied through leaky windows or special registers located either in the wall or in the window frame. Nevertheless, for the system operation as described here, all dwellings must be isolated from staircases or elevator shafts.

Because the air pressure difference varies for each floor, the airflow rate will depend on the space location. In an actual building, pressure disparity will be partly compensated by the airflow resistance of the ventilation shaft. Nevertheless, space on the last floor will need to be better ventilated, especially when the roof is flat.

Effectively, one must design all buildings, small and tall, to achieve a pattern shown in Figure 3.1. All floors must be separated with airtight floors, and all connections from the ground to the staircases, elevators, or other vertical shafts must be reduced to a minimum. Furthermore, it is critical that each dwelling is directly connected to the outdoors on the same level. In other words, the design of a high-rise is like a design of several independent bungalows, stacked one over the other, and directly connected to the outdoor air.

Another driving force of air exchange is wind (Figure 3.4) and eq. (3.12):

$$\Delta p_w = K \frac{v^2 \rho_e}{2} \tag{3.3}$$

where K is the aerodynamic coefficient, v is the wind speed, and $\rho_e$ is the exterior air density.

Wind pressure on the windward wall will drive air into the building by infiltration through windows, gaps, and cracks in the external shell. On the other side, at the leeward wall, the under-pressure of air will draw air out of the building through leaky windows and walls. As a result, a horizontal, internal air exchange will be observed in the building. Windward spaces will be intensively ventilated with fresh air, while stale air will be supplied to downwind areas.

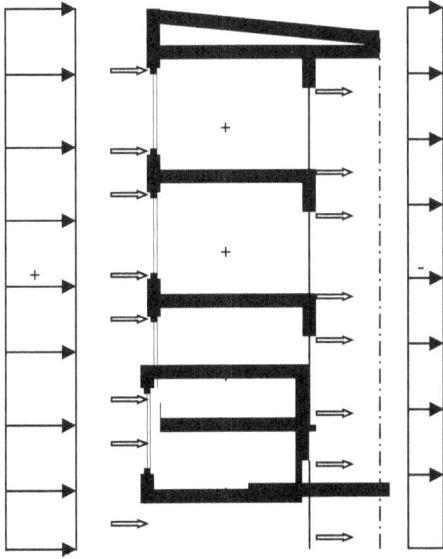

**Figure 3.4:** Wind-driven air exchange.

In an actual building, a complicated and dynamic superposition of different effects will take place. Caused by different forces and pressure modulation from air conditioning and ventilating devices, an unpredicted air exchange may occur, changing the actual fresh air delivery. The following factors may influence natural ventilation intensity:

– Building shape, orientation, wind speed and direction, other buildings, and sheltering trees
– External and internal temperature differences, including solar radiation
– Local changes in the air tightness of the building enclosure and the vertical position of the dwelling
– Horizontal location on a floor in relation to vertical shafts or staircases
– Air tightness of the building entrance doors and the penthouse or mechanical room on the roof

Still, as an overpressure is used in mechanical ventilation to deliver direct outdoor air to the indoor space, one may use an on/off ventilation scheme for exhaust openings to invoke hybrid ventilation and ensure good ventilation while keeping the average air pressure close to the ambient. If one ensures, by other means, no risk of moisture entry into walls, one could also operate an overpressure ventilation.

## 3.4 Development of the current approach

Airflow impacts all environmental transport and, thereby, the indoor environment and durability. To a lesser degree, the contamination of wall cavities by organic materials carried by the airflow from inside or outside the building provides both the nutrients and spores for mold growth.

### 3.4.1 Terminology related to air barriers

We need to distinguish between different concepts of air tightness, namely:
1. Material airtightness is measured in the laboratory; *we will not use it in practice.*
2. Joints airtightness. Airflow per lineal perimeter or window. See point 5a.
3. Clear wall airtightness, measured in the laboratory on an 8 ft × 8 ft plain wall (NRC 1988, 1990; MH, 1991; see HRef). The typical criterion is set at 0.20 L/(s · m$^2$).
4. Envelope airtightness of the whole building enclosure, measured under field conditions, is expected to pass the criterion of 2.00 L/(s · m$^2$).
5. Envelope section airtightness can be determined in two ways:
   (a) Measured directly, for example, a box attached to the tested window, a few blower doors.
   (b) Calculated from multizonal network with airflow perturbations (Lstiburek, 1998).

Laboratory tests are for manufacturers, but do not define the field performance. The latter depends on many pathways between different zones, air flows through interior partitions, and the dynamic response of the enclosure, different from the laboratory tests. *Quiruette* (1985, see HRef) published field measurements of a brick veneer wood-frame wall, where under dynamic wind gusts, 70% of the air pressure reduction took place on the brick veneer layer, while in the laboratory, testing performed with static air pressure showed that 70% of the air pressure drop recorded on the polyethylene material of the air barrier system and 30% on the brick veneer.

### 3.4.2 Background to current requirements of airtightness

Wilson and Novak (1959, HRef) showed that, in the worst case, the vapor transfer by air leakage was ten times larger than that gained by diffusion. There was a widespread publication of similar findings (Wilson, 1960; Thorpe and Graee, 1961; Sasaki and Wilson, 1962, 1965; Garden, 1965; Wilson and Garden, 1965, see HRef), document-ing the significance of airflow in moving water across or within the building enclo-sure. Despite a significant number of publications that stressed the need for air leak-age control (Wilson, 1960b, 1961; Tamura and Wilson, 1963, 1966, 1967; Garden, 1965, see HRef), airtightness, being related to the quality of construction, has been largely ignored. A change in emphasis only came when the field data confirmed that humid-ity carried by air is a significant source of condensation.

The energy crises of the mid-1970s and passive house demonstrations motivated Canadian builders to switch to electric heating. Eliminating combustion flues, how-ever, reduced air exchange and increased humidity in these houses. Condensation problems in attics became more frequent (Stricker, 1975; Orr, 1974, HRef) and in flat wood-frame roofs (Tamura et al., 1974, HRef). Several studies (Wilson, 1960; Tamura and Wilson, 1963; Tamura, 1975, HRef) showed that changes in the position of the neu-tral pressure plane caused changes in ventilation efficiency.

Furthermore, the elimination of chimneys increased air pressure in the upper part of the room and resulted in air and water-vapor flow into cold attics and water accumulation there (Dickens and Hutcheon, 1965, HRef). Elmer and Levin (1983) recognized that air movement from conditioned spaces to unconditioned attic spaces resulted in water condensation in attics, led to the recommendation that the air tightness of the ceiling and partition-to-ceiling details needed to be im-proved.

The reduced ventilation rate in the electrically heated houses created several cases with so-called building syndrome, that is, houses with zones or areas with insufficient ventilation rates. To eliminate both problems, mechanical ventilation was necessary. The model building code of Canada in 1985 required dwellings to have a mechanical ven-tilation system. The introduction of mechanical ventilation set the stage for further im-provements in airtightness. A consensus has developed that is critical to the success of buildings is an approach (CHBA, 1989 Href) that integrates energy efficiency and a healthy indoor environment. To this end, new design tools for integration and enhanced tradeoffs between the equipment and building enclosure must be considered.

The following were the critical technical requirements used in both the Canadian R-2000 program and the Building America Programs:
- Use of mechanical ventilation
- Follow specific requirements for an air barrier system
- Design to avoid thermal bridges
- Control of moisture entry from the ground
- Perform voluntary testing for balancing the ventilation systems

Two approaches to air leakage control were introduced earlier – the Airtight Drywall Approach (ADA) and the External Airtight Sheathing Element (EASE). The ADA system was developed by Lstiburek and Lischkoff (1984, HRef). Using gaskets and controlling the terminations of the drywall sheets, they achieved relatively airtight buildings.

The vapor resistance was provided using paint on the drywall. The measures to achieve air tightness involved using both polyurethane foam and neoprene gaskets at the terminations of the drywall.

While ADA systems have been shown to work well in single-family houses, they did not satisfy the needed sound transmission control in row housing or apartment blocks.

The EASE, with a continuous layer of thermal insulation on the exterior, reduced the effect of thermal bridging. It reduced the risk of condensation in the wall cavity by increasing the temperature on the surface facing the wall cavity in the wood-frame wall. Recently, the minimum R5 continuous exterior insulation requirement has been added to the Canadian Building Code.

In the 1995 edition of the NBC of Canada, the air barrier system was required to satisfy the following requirements:

–   There should be a continuous material layer to provide the principal resistance to air leakage. The air barrier material permeance was not greater than 0.02 L/$(s \cdot m^2)$ at a 75 Pa difference, but we do not support this number.
–   All components of the air barrier system shall comply with the durability requirements specified by the respective material standards.
–   The system shall be continuous across joints, junctions, and penetrations.
–   The system shall be capable of transferring design wind loads, and
–   The system shall be evaluated with deflections 1.5 times the specified wind load.

### 3.4.3 The need to test air flow rates

Despite its significance, quantifying the effects of airflow has yet to be improved. The amount of air moving into a building is determined by the air flow resistance of the building envelope and the pressure difference between the inside and outside of the building. The difference across the building enclosure is created by wind and temperature differences, heating/cooling, and other mechanical devices used by occupants.

The airtightness can vary throughout the building, with different wall sections, with or without windows, corner sections, with or without doors, etc. The air pressure depends on wind speed, direction, and air temperature differences related to the building's shape and dimensions. The typical field-testing methodology has been focused on air leakage of the whole building, and the relation between laboratory airtightness of a wall assembly (clear wall) and that of the assembly in the building is, at best, tenuous.

Airtightness of building enclosures is needed for the following reasons:

–   To control the air exchange rate, since it affects *energy use*
–   To reduce the amount of water carried by air to improve the *durability* of a structure
–   To *maintain a healthy indoor environment* by reducing the mass of VOCs stand for Volatile Organic Compounds. These are a group of organic chemicals that readily evaporate at room temperature and can significantly contribute to indoor air pollution. Common examples include formaldehyde, benzene, and toluene, which are

often emitted from building materials, paints, adhesives, cleaning products, and other household items, particulates, or mold spores carried to the indoor space
- To reduce the heat gained or lost when the temperature of the entering air differs from that in the indoor space
- To control and balance air exchange between separate zones or rooms

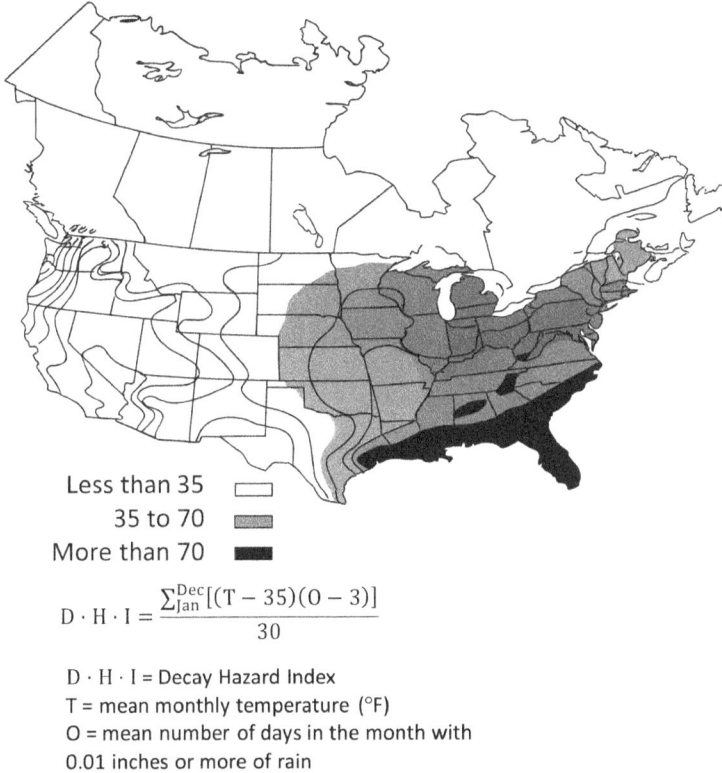

Less than 35
35 to 70
More than 70

$$D \cdot H \cdot I = \frac{\sum_{Jan}^{Dec}[(T - 35)(O - 3)]}{30}$$

D · H · I = Decay Hazard Index
T = mean monthly temperature (°F)
O = mean number of days in the month with
0.01 inches or more of rain

**Figure 3.5:** Amount of moisture deposited (kg/m²) in different climates (Bomberg et al., 2016).

Figure 3.5 shows the amount of moisture (kg/m²) condensed from the flow of air with a temperature of 21 °C and 35% RH entering the wall at the rate of 0.9 L/(m²·s·Pa) at a 50 Pa pressure difference in different locations as a function of season. In some cases, energy efficiency can outweigh durability; in other cases, the opposite will be correct. Moisture accumulation over the year in locations like Toronto or Helsinki is on the borderline of an unacceptable yearly increase, while Vancouver is far from such a risk. Moisture accumulation depends not only on wetting but also on drying conditions.

Figure 3.6 shows a map of warm temperatures and heavy rainfall in the USA that indicates regions with a high potential for wood decay. Similarly, in Vancouver, the rainy season coincides with mild winters, and the hygroscopic nature of the wood makes the decay potential for wood in Vancouver higher than that in Toronto. Yet, whether the higher potential leads to actual decay depends not only on the drying rate but also on the water storage capacity of the wall assembly.

This example illustrates the complex nature of durability assessment. One cannot predict the outcome of complex phenomena based on one part only, however critical it may be. Furthermore, moisture is only one part of the environmental complex. We must address heat, air, and moisture transport simultaneously because of the strong interactions between the three phenomena.

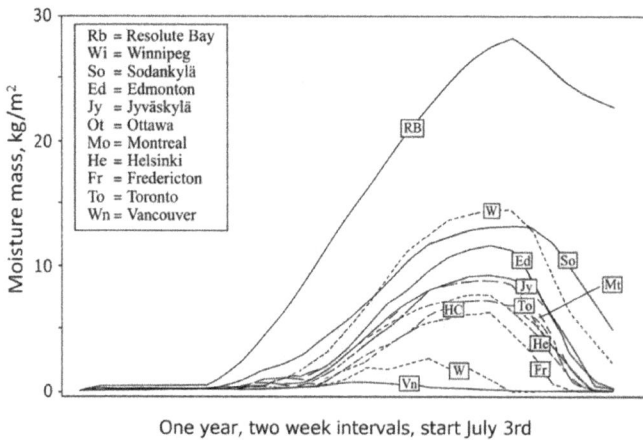

One year, two week intervals, start July 3rd

**Figure 3.6:** A decay potential for wood in North America (Bomberg et al., 2016).

### 3.4.4 Different air pressure fields

Now, we have well-insulated, airtight buildings in which there is a potential for increased health problems caused by mold or microbial contamination if the system does not work correctly. Today, air pressure fields can significantly affect building enclosure performance. Understanding air movement in buildings is necessary for successful design. The determination of air pressure differences, however small and challenging to measure, is needed to establish the performance of the building as a system.

A strategy to control air pressure in the building space includes the following steps:
1. Enclose the air space and quantify the degree of airtightness
2. Use controlled mechanical ventilation
3. Control air-pressure fluctuations induced by heating, ventilation, and air conditioning (HVAC) system operation
4. Control air-pressure gradients induced by HVAC system operation
5. Eliminate interconnected internal cavities communicating with the HVAC system.

6. Review the building's mezzo-climate for differences in wind and solar shading conditions.

The effect of pathways created by external cavities and interconnected internal cavities communicating with HVAC systems on the performance of building systems should be addressed. The significance of these elements has been illustrated in the few examples (Figures 3.8 and 3.9) selected from case studies (Lstiburek, 1992, 1994, 2000, see HRef; Lstiburek, 1998; Lstiburek et al. 2002).

- A facility in a cold climate with inadequate provision for return air. When the interior doors were closed, individual spaces became pressurized. The common areas, in turn, become depressurized. When atmospherically vented combustion appliances (such as fireplaces and gas water heaters) were located in the common areas, the negative pressure in these regions led to the back-drafting of combustion appliances. In pressurized spaces, the forced exfiltration of interior (typically moisture-laden) air led to condensation and moisture-induced deterioration problems.
- Hallways and corridors can cause an extension of pressure fields throughout a building (Figure 3.7). A typical hotel room ventilation system may have a bathroom exhaust operating continuously via a rooftop-mounted fan (which also serves other bathrooms). Make-up air for this bathroom exhaust is typically provided through the exterior wall via a unit ventilator or packaged terminal heat pump (PTHP). In the case investigated, the design assumed that 60 cfm out through the bathroom was offset by 60 cfm through the unit ventilator or PTHP. Although the unit ventila-

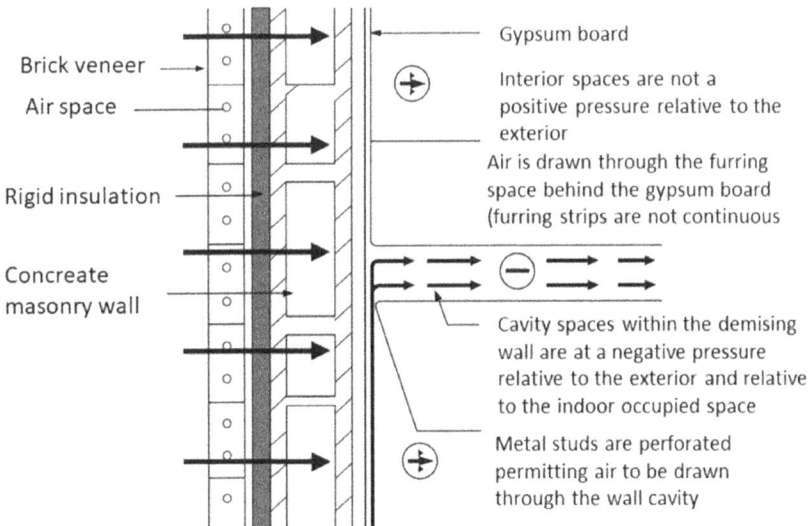

**Figure 3.7:** Connection through the furring space extends exterior air impact; despite the positive pressure in the interior, mold grows on interior walls (Bomberg et al., 2016).

tor or PTHP did not run continuously, an intermittent imbalance of 60 cfm was not considered a problem.

Now consider 30 hotel rooms on a floor served by a single corridor which connects rooms on the floor. With 30 exhaust flows of 60 cfm each, if they operate continuously, an 1,800-cfm exhaust is created on the floor. Unit ventilators or PTHPs only operate on a 20% duty cycle (i.e., 80% of the units are not operating at a given time). The air supply was only six operating unit ventilators of PTHPs at 60 cfm each, and the flow imbalance was 1,440 cfm. In hotel, facilities in hot and humid climates, the negative pressure field is the single most significant reason for the mold, odor, and moisture damage.

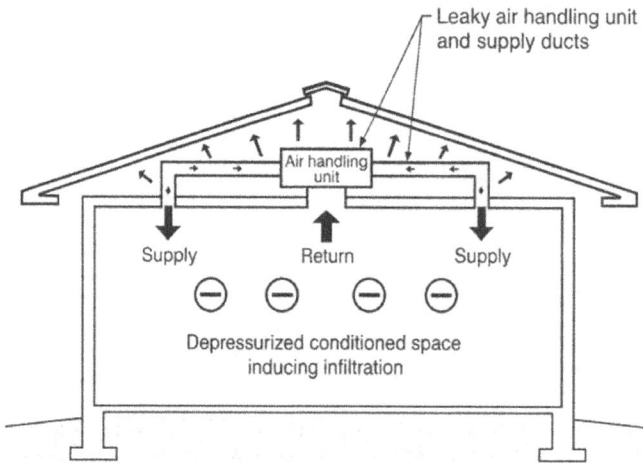

**Figure 3.8:** A facility located in a hot, humid climate with leaky ducts located outside the conditioned space. Air leakage depressurized the conditioned space, causing an infiltration of exterior hot, humid air (Bomberg et al., 2016).

Figures 3.8 and 3.9 highlight that air leakage-pressure relationships are the keys to understanding the interaction between the building enclosure and the HVAC system. We must control pressure fields to design and build safe, healthy, durable, comfortable, and economical buildings.

## 3.5 Air leakage of a building

The typical relation between the air leakage and the pressure difference between the building and its environment is expressed as follows:

$$Q_a = C \, \Delta p^\beta \tag{3.4}$$

**Figure 3.9:** Tall building in Central Europe in winter. Air exfiltration causes frost to be visible on windows of the top floors, while those at the low levels are clear. Photo: Bomberg, 1991.

where C and β are experimental coefficients. The pressure exponent is typically found in the vicinity of 0.65 but has limiting values of 0.5 for a turbulent flow and 1.0 for a laminar flow.

The power law is a reasonably good empirical description of the flow-versus-pressure relationship; however, it does not correspond to any physical paradigm. There are physical paradigms that could be applied to the problem of airtightness:

– If the flow rate (Reynolds number) is low enough, the flow will be dominated by laminar frictional losses, and the flow will vary linearly with the pressure drop.
– If the leak is small, frictional forces in the leak itself can be ignored, and the leak may be treated as an orifice in which the flow is proportional to the square root of the pressure drop. The higher the flow rate (i.e., Reynolds number), the leak can still be treated as an orifice.

There is no direct relation between airtightness and ventilation. One is related to the quality of construction and dealing with the building enclosures. It is expressed in $L/(m^2 \cdot s)$ or in $m^3/(m^2 \cdot h)$, measured at 50 Pa. The ventilation is expressed as air change per hour (ACH). The first measure relates to the area of the enclosure, and the second relates to the floor area and the number of occupants, because the level of ventilation depends on the degree of air pollution. Now the question becomes: can these two phenomena be correlated?

Requirements for air tightness vary, but typically in UK one requires a building to have air leakage below 3 $m^3/(m^2 \cdot h)$ when tested with a 50 Pa pressure difference. For a typical Canadian bungalow with 110 $m^2$ in floor area, 440 $m^3$ of heated volume, and 220 $m^2$ of the envelope surface area, this would give a leakage rate of ($3 \times 110 = 660$ $m^3/h$), and with a wall area of 440 $m^2$, we get 1.5 $m^3/h$ or 1,500 $L/m^2h$. Assume a value of $\beta = 0.65$ and recalculate this requirement for $\Delta P$ of 4 Pa, we get the result ~0.33 $L/m^2s$. Thus, using the rule of thumb, we see that international practice sets the minimum requirements close to the minimum for fresh air needed for breathing.

For residential occupancy, one may require mechanical ventilation conforming to the rates required by ASHRAE Standard 62.2 – 2013, "Ventilation for Acceptable Indoor Air Quality." Air tightness requirements for the building enclosure are confusing because of three levels of air leakage control:
1) Level of material, the so-called "plane of air tightness in the cross-section," or the principal resistance to air leakage through materials.
2) Level of the wall assembly, with joints between material sheets, for example, sheathing boards. This level would typically include the connection between the wall and the window.
3) Level of a building with staircases and ducts, and frequent use of a return plenum joining different indoor spaces in office buildings.

Details of joints between continuous air barrier materials that make the large difference on material level, presence of low-density polyurethane foam that does not qualify as material for air barrier system but works in the practice as such, raise the question of the need for a material to have an air flow lower than 0.02 $L/(m^2 \cdot s)$, and we will not use this criterion.

The second problem is created by measuring field air leakage at 50 Pa air pressure, while the range of 4–10 Pa applies to the operating conditions. There is some agreement between the recommended criteria and the practice of low-energy, low-rise houses tested in 1980s in Saskatchewan *(BRN 178)* with 1.45 exchange per hour and R-2000 houses in Canada that varied from 1.15 to 1.35 ACH (Canmet, 1997, HRef). Observe, however, that the criterion is close to minimum needed for health, and in the case of electrical failure, people will be forced to open windows even during inclined weather.

The "standard" multiunit dwellings in the USA and Canada have much lower airtightness. Several measurements performed on large apartment buildings (Shaw et al., 1991) with fan depressurization technique showed values 3.2 to 5.2 $L/(m^2 \cdot s)$ at

$\Delta P = 50$ Pa for one building and 2.3 L/(m²·s) to 3.6 L/(m²·s) for three other buildings (5 to 17 stories high).

In effect, realizing that the air leakage also depends on the pressure distribution in the building space and many other factors related to the building HVAC and occupancy, we do not propose any specific criterion except for a benchmark to force builders to perform an air leakage test as a means to quality assurance.

## 3.6 Methods for airtightness testing under field conditions

Traditional airtightness tests are suitable for the measurement of a single zone, where the entire space is uniformly pressurized or depressurized. Blower-door technology was first used in Sweden around 1977 as a window-mounted fan (Kronvall, 1980, HRef) to test the tightness of building envelopes (Blomsterberg, 1977, HRef). The same technology was pursued by others (Caffey, 1979; Harrje, Blomsterberg, and Persily, 1979, HRef) to help find and fix leaks (Sherman, 2004, HRef).

### 3.6.1  Fan depressurization method EN-ISO 99722

If the air leakage returns the quantity of air exhausted by the blower as a function of the measured pressure, as shown in eq. (3.4), the two coefficients can be determined. One makes a series of measurements to increase the precision of the value estimated at 50 Pa air pressure difference (Figure 3.10). Using this methodology, one may quickly test small houses at different times and locations to examine how airtightness changes over the seasons and how builders in each location respond to climate requirements. Indeed, Figure 3.11 shows five provinces and national averages over a span from 1981 to 1995 (Proskow, 1998).

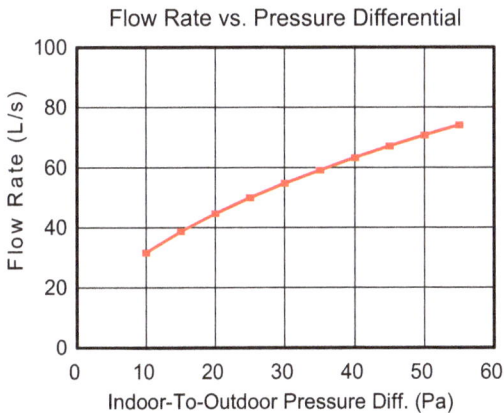

**Figure 3.10:** Typical results of blower-door test.

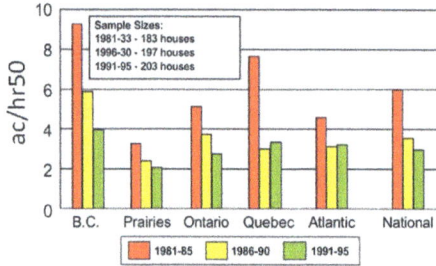

**Figure 3.11:** Airtightness of houses in Canada measured from 1981 to 1995 (Proskiw, 1998).

The effect of climate on air tightness is clear. In the 1990s, in the mixed climate of British Columbia, the average was four ACH, and in the cold climate in Quebec and Ontario, the average was near three ACH. Finally, in the severe cold of the Prairies, the average airtightness was two ACH. In small buildings, one used only one blower door. Testing large buildings may require more blower door types of equipment. Shaw (1981, HRef) used three blower-door setups to compensate air flows on the adjacent floors to ensure one-dimensional flow of air.

This technique is difficult because of randomly varying conditions on each floor; it takes time to adjust three blowers to identical conditions at three locations. Furthermore, this method does not eliminate the effects of penetrations located between the zones.

### 3.6.2 Parallel flow airtightness test

Proskiw (1998) used two blower doors in the parallel flow airtightness test. This allows measurements of air transfer between Part A and Part B of the multizonal building. This procedure consists of three steps:

**Step 1:** Install the blower door between zones A (closed) and B, and open zone B to the outside. Measure the flow from zone A to B.

**Step 2:** Install the blower door in zone B and exhaust it to the outside.

Next, switch off the blower door in zone B and start the one in zone A. One has measured pressure differences from zone B to the outdoors and between zones A and B.

These measurements are considered as the initial set of conditions that give $\Delta P_A$ and $\Delta P_B$.

**Step 3:** Adjust the blower door in zone A to measure the pressure difference between zones A and B, the same as measured in stage 2 as $\Delta P_A$.

**Step 4:** With the air leakage across the exterior wall being the same as measured in step 2, one can write a series of equations describing the difference in the airflow for the partition between A and B as follows:

$$Q_T = Cp \left( \Delta P^n_{binitial} - \Delta P^n_{bfinal} \right) \tag{3.5}$$

where Cp is the specific heat of air.

Because we have two tests with the blower door number 2 (B), one can solve the flow equation. As discussed in Section 3.6, the coefficient n must fall in the interval 0.5–1.0, or the test is discarded. This method was verified in the laboratory and field application (Hult et al., 2012–2014).

### 3.6.3 Using perturbations and inverse of multizonal flow solutions

While two-blower solutions are very useful for assessing flows through two parts of the building separated by doors; however, one may also need to address a specific part of the building enclosure, for example, a room where people complain about the presence of odors. Lstiburek (1998) used a perturbation technique to solve several multizonal flow equations with a blower door and an interzonal air flow model, for example, CONTAM. The perturbations were created by opening and closing windows and doors. Yet, an interzonal flow model uses air pressure as an input, and our problem is to find values of pressure; therefore, one must guess the solution and try the model to see if the solution yields reasonable flows. This makes the unknown uncertainty in the inverse solution of the multizonal flow system.

### 3.6.4 Gas tracer method

Discussed so far, pressurization methods describe air flow at conditions different than actual operating conditions in the building that is both advantage and a disadvantage. The advantage is that it is mostly independent of weather conditions and results in enable building tightness rating.

The disadvantage is that it gives no information about actual ventilation capability. The opposite is true for the gas tracing method that allows the determination of instantaneous flows in relation to actual interior and exterior conditions (temperature and wind). The measurement method is based on either multiple or continuous tracking of the concentration of a gas introduced into a specified building space. The simplest measurement approach is the concentration decay method, where tracer gas is supplied once to the tested space, and the air exchange rate is evaluated from the concentration decay. The procedure is given in an EN-ISO standard (EN-ISO 12569):

$$C(t) = C_0 \cdot e^{-nt} \tag{3.6}$$

Where C is the concentration as a function of time, $C_0$ is the initial concentration, and n is the air infiltration rate (1/h). Knowing the infiltration rate, one can calculate the air exchange rate (ACH) for a specified volume.

This method assumes that the mixture of air and a tracer gas behaves as an ideal gas. To ensure good measurements with the tracer gas, we require that it has a density close to the air. If the tracer gas is not mechanically mixed with air, measurements in the initial period may not be used for calculations. Furthermore, one must verify whether the self-mixing of air with the tracer gas is not changing the internal temperature distribution and/or stratification.

Another problem is connected to uneven trace gas concentration in the whole space volume, especially in big spaces or complicated building geometry. This problem can be dealt with:

–   Air samples were collected at several points to establish the path of the flow (Sherman at al 2011a, 2011b, 2014).
–   Measurements are conducted at several points near the location of air escaping the building, and the average value is used when calculating the ventilation rate (Elmroth and Levin, 1983). In addition to the one presented above, two other tracer gas methods can be used (EN-ISO 12569):
–   Continuous dose method.
–   Constant concentration method, or their combinations.

Selection of the method depends on a building structure, the kind of ventilation system, building use, tracer gas, detection system availability, and cost. In the case of multicell spaces, the constant gas concentration method is the only one that may yield reliable results because it is not affected by cross-flows of air between building cells or zones.

A wide choice of tracer gases is available now, each characterized by an individual set of advantages and restrictions (EN-ISO 12569): helium, carbon dioxide, sulfur hexafluoride, perfluorocarbon, ethylene, and nitrogen monoxide. $CO_2$ seems to be the most common, easy-to-track, and harmless substance, considering its global warming potential or flammability, and when measurement does not require high accuracy. The main problem with $CO_2$ as a tracer gas is that it is usually generated in buildings by the occupants and other sources. $CO_2$ generation rate must be known to measure its concentration changes due to air leakage.

## 3.7 Discussion on air control issues

Current trends toward sustainable buildings require a fundamental revision to many design premises developed over the years. The most important change is the integration of design and construction, quality assurance, and commissioning processes. The interaction between mechanical systems and building envelopes became significant and impacts every aspect of the built environment. Issues such as indoor air quality, fire protection (smoke, toxicity, and fire spread), durability (moisture accumulation), comfort (temperature, relative humidity, odors), and the cost of operation and maintenance are strongly related to airflow control. The advent of low-energy buildings (near zero or net zero energy) has brought us the interacting design of the building enclosure and HVAC systems because these systems affect wetting and drying of walls, rain penetration, pollutant migration, and the durability of the building enve- lope. All these phenomena are centered on heat, air, and moisture movements that we label environmental control. Controlling air movements is the most important of these three means of transport. To highlight it, Lstiburek (1998) introduced the term "air pressure response of the building."

Air pressure differences are small, typically in the 0.5 to 2 Pascal range, but they significantly affect building performance. Lstiburek (1998) stated:

> In most mid-rise and high-rise buildings, the stack effect typically dominates the HVAC system air flows. . . . Air flows from the lower units and floors, up the elevator shafts, stairwells, and service penetrations to the upper branches and floors. This stack effect-induced air flows are re- sponsible for pollutant migration, odor problems, smoke and fire spread, elevator door closure problems, and high thermal operating costs. By sealing units from corridors and by isolating cor- ridors from elevator vestibules, stack effects are reduced, and the pressure drops occur across the corridors and elevator vestibules, not the exterior building envelope. This results in safer buildings concerning smoke and fire control. Indoor air quality problems are reduced, and en- ergy efficiency is greatly enhanced.

While air transport control is now recognized as a critical issue in building enclosures design, achieving a balance between energy efficiency, indoor environment, and durabil- ity concerns is still a challenge. One needs to review the hygrothermal performance of the whole building simultaneously as one examines the building enclosure. As the distri- bution of leakage areas and air pressure relationships from wind and stack forces can- not be determined, to ensure indoor air quality, health, and safety, one must use me- chanical ventilation in all buildings regardless of building envelope tightness.

Ventilation serves two purposes: (1) source control for removing pollutants, for example, exhaust in the kitchen and bathroom, and (2) whole house ventilation to di- lute and remove other pollutants. Typically, the whole house ventilation works inter- mittently, defined as a fraction of each hour during occupancy needed to provide the required amount of fresh air. Furthermore, the air distribution system ensures deliv- ery or exhaust (or both) occurs in each room. Suppose the building is provided with source control. In this case, mixing can be introduced either by the interior airflow organization, for example, supply air to the bedrooms and exhaust from the kitchen, or through a bypass on the heat recovery ventilator. The latter may be preferred for periodically used spaces and requires advanced controllers. For the equivalency of intermittent ventilation and dynamic controls, see Turner and Walker (2012, 2013) and Sherman and Walker (2011a, b).

Finally, we must come up with recommendations for the range of airtightness for contemporary buildings. Over the years, many works reviewed the relationship be- tween building enclosure airtightness and air ventilation requirements, starting with Shaw (1981) to Chan et al. (2011, 2012a, 2012b) and Walker et al. (2013), where the latter works are based on more than 140,000 homes database. We also know that designing low-energy buildings (now called **zero-energy ready** buildings), we cannot use one template for all climates. To design a zero-energy building, we need to consider peak loads (unless they are equalized, e.g., as in PTAC technology) and the average energy use in heating and cooling; all are climate-related. Airtightness must also be adequate to satisfy moisture-originated durability considerations. Furthermore, as air leakage relates to the area of exterior walls, it relates to the building's shape and size.

For small buildings in a cold climate, the benchmark of the building enclosure is $3 \ m^3/(m^2 \cdot h)$ or $0.8 \ L/(m^2 \ s)$ or 1.5 ACH. One can postulate 3 ACH at 50 Pa for larger buildings in milder climates. Testing airtightness during the construction process allows improving the construction quality; testing completed construction means missing the opportunity for improvement. Since the wall-floor, wall-window, and wall-roof connections are the typical locations of faults, we recommend using pressure boxes for small components combined with a blower door and infrared testing. Thermal imaging combined with significant pressure differences during blower door tests is also useful.

Thermal image shown in Figure 3.12 was taken during a very hot period. Because of the efficient thermal insulation of the walls and windows, high air tightness, and massive structure, the indoor temperature in the tested building was 11 K lower than the ambient temperature. Using a combined blower door and infrared test of a passive school building ($\Delta p = -100$ Pa, $T_e = 32$ °C, $T_i = 21$ °C), warm external air leakage was observed (Bomberg et al., 2016).

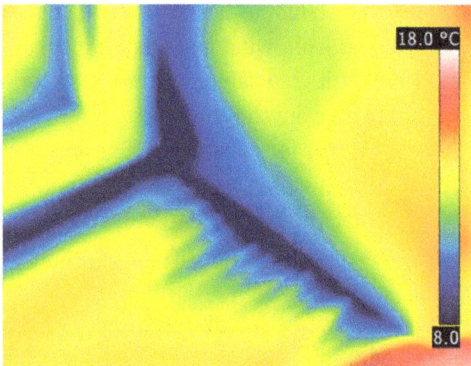

**Figure 3.12:** Thermal image taken during a hot weather period (Bomberg at al., 2016).

## References

Bomberg, M.; Kisilewicz, T.; Mattock, C. Methods of Building Physics, Cracow University Press, pp.1–300. also free acess on the internet, 2016.

Chan, W.R. Sherman No, 4, pp M H. 2011, Preliminary Analysis of U.S. Residential Air Leakage Database v.2011. *2011 32nd AIVC Conference and 1st TightVent Conference: Towards Optimal Airtightness Performance*, Download: PDF 1.66 MB)

Chan, W.R.; Joh, J.; Sherman, M.H. *Air Leakage of US Homes: Regression Analysis and Improvements from Retrofit.*, 2012a. Download: PDF (1.88 MB).

Chan, W.R.; Joh, J.; Sherman, M.H. *Analysis of Air Leakage Measurements from Residential Diagnostics Database*. 2012b. Download: PDF (2.44 MB).

Elmroth, A.; Levin, P. Air Infiltration Control in Housing – A Guide to International Practice, Swedish Council fo Buiklding Research. 1983, ISBN 91–540-3853-7, D2.

EN-ISO 13829 Thermal performance of buildings – Determination of air permeability of buildings – Fan pressurization method.

EN-ISO 12569 Thermal performance of buildings and materials – Determination of specific airflow rate in buildings – Tracer gas dilution method.

Hult, E.; Sherman, M.H.; Walker, I.S. Blower door techniques for measuring interzonal leakage. In Proc. Thermal Performance of the Exterior Env. of Buildings XII, ASHRAE/DOE/BTECC, 2013.

Hult, E.L.; Sherman, M.H. Estimates of Uncertainty in multi-zoned air leakage measurements. *Int. J. Vent.* 2014, Download: pdf.

Hult, E.L.; Dickerhoff, D.J.; Price; P.N. *Measurement Methods to Determine Air Leakage Between Adjacent Zones*. 2012, Download: PDF

Lstiburek, J.W. Toward an Understanding and Prediction of Air Flow in Buildings, PhD Thesis, 1998, U. of Toronto.

Lstiburek, J.W.; Pressnail, K.; Timusk, J. Evaluating the Air Pressure Response of Multizonal Buildings. *J. Build. Phys.* 2002, *25*, 299–320.

Proskiw, G. Parallel flow airtightness test (PFAT), from M.Sc. thesis at Concordia U, 1998, Montreal,

Shaw, C.Y. A correlation between air infiltration and air tightness for houses in a developed residential area. *ASHRAE Trans*. 1981, 1981, *87*, Pt 2 pp. 333–341.

Sherman, M.H.; Walker, I.S.; Lunden, M.L. Uncertainties in Air Exchange using Continuous-Injections, Long-Term Sampling Tracer-Gas Methods. *Int. J. Vent.* 2014, *13*(1), June 2014, Download: PDF.

Sherman, M.H.; Walker, I.S.; Logue, J.M. Equivalence in ventilation and indoor air quality. *HVAC&R Res.* 2011, 2011. Download: PDF.

Sherman, M.H.; Walker, I.S. Meeting residential ventilation standards through dynamic control of ventilation systems. *Energy Build*. 2011, *43*, 1904–1912. Elsevier. LBNL-4591E.

Turner, W.J.N.; Walker, I.S. Advanced Controls and Sustainable Systems for Residential Ventilation. 2012 Download: PDF.

Turner, W.J.N.; Walker, I.S. Using a ventilation controller to optimise residential passive ventilation for energy and indoor air quality. *Build Environ*. 2013, 70(2013), 20–30.

Wallburger, A.T.; Bomberg, M.; Henderson, H. Energy prediction and monitoring in a High-Performance Syracuse house. 2010, Available at: http://thebestconference.org/BEST2

Walker, I.S.; Sherman, M.H.; Joh, J.; Chan, W.R. Applying large datasets to developing a better understanding of air leakage measurement in homes. *Int. J. Vent.* 2013 2013, *11*(4), 323–338. Available at: http://www.ijovent.org/doi/abs/10.5555/2044-4044-11.4.323.

# Chapter 4
# Interacting water and vapor flows

A term "moisture" denotes the presence of water in an undefined phase. Understanding interactions between air, water, vapor, and ice, if such exists in pores of materials, is critical to material science. Therefore, we will start by considering the vapor phase in the air and proceed to the liquid and vapor in the material's pores.

## 4.1 Water vapor in the air

Total gas pressure equals the sum of different component partial pressures (p). Each of the gases tends to equalize its partial pressure (Dalton's law). For each mole of the gas

$$p V = (m/M) R T \qquad (4.1)$$

where V is volume [m³], m is mass of the gas [kg], M is molecular mass of the gas [kg/mol], R is 8.314 [J/(kmol·K)], and T is absolute temperature [K].

The molecular mass of dry air = 28.96 [kg/kmol], and18.02 [kg/kmol] for water vapor.

So, the density of dry air is $\rho = m/V$, or

$$\rho = (p\,M) / (R\,T) = p / (240.8\,T), \quad \left[ kg/m^3 \right] \qquad (4.2)$$

In ventilation technology, water vapor content in kg per kg of dry air is used.

$x = m_v / (m_{tot} - m_v)$, or if using partial water vapor pressure:

$$x = 0.622\, p_v / (p_{tot} - p_v) \qquad (4.3)$$

Alternatively, the concentration of water vapor is divided by the density of dry air,

$$x = c / \rho_a \qquad (4.4)$$

The relation between temperature and phase change of water is shown in Figure 4.1.

In this book, we do not discuss psychrometry, except for the concept of relative air humidity. The relative humidity is the ratio of its partial pressure to the saturated partial pressure of water vapor at a given temperature. Figure 4.2 highlights the relationship between the moisture storage of air and temperature.

At a temperature of 20 °C, the maximum vapor concentration is 17.3 g/m³, so 8.65 g/m³ of WV in the air corresponds to 50% RH; at 10 °C, the same absolute humidity gives 92% RH, and for 5 °C, condensation has already occurred, and the remainder of water vapor is in equilibrium with the liquid phase. We consider moisture equilibrium as if it was a static phenomenon, while it is a dynamic process of water continually

https://doi.org/10.1515/9783112217023-004

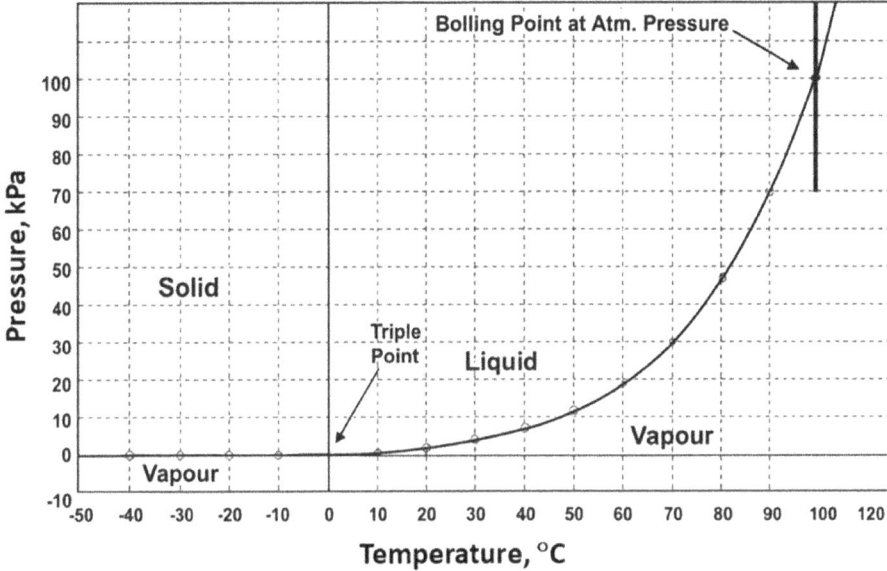

**Figure 4.1:** State diagram for water shows the equilibrium between liquid, vapor, and ice.

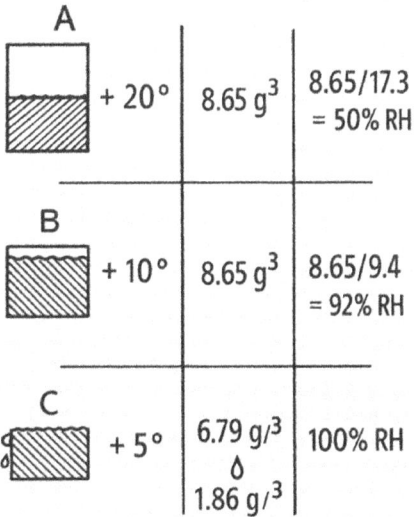

**Figure 4.2:** The same amount of moisture in indoor air (8.65 g/m$^3$) results in different relative humidity values when the temperature falls from 20 °C to 10 °C or 5 °C.

evaporating and condensing on the liquid surface. If the rate of condensation and evaporation are equal, we may talk about a macroscopic equilibrium. The dynamic nature of all transport phenomena is characterized by their statistical nature.

Water vapor contained in indoor air comes from a few sources, namely:

1.  Ventilation air brings water vapor from outdoors.
2.  Unplanned air flows (UAFs), that is, exchange between outdoor and indoor air that acts parallel to a ventilation.
3.  Moisture production by occupants (e.g., in kitchen or bathroom).
4.  Sinks or sources of moisture stored in furniture, books, and finishing materials.

Disregarding (2) and (4), we may consider the water vapor concentration in indoor air ($v_L$, kg/m$^3$) as that of the outdoor air ($v_o$, kg/m$^3$), plus moisture production by occupants (G, kg/h), and the product of the air volume (V, m$^3$) and the ventilation rate per hour (n, 1/h). Term $G/(n·V)$ is called the moisture product. If, at time $t = 0$, the indoor concentration is equal to the outdoor concentration, namely:

$$V_L = v_o + G/(n \cdot V) \; \{1 - \exp(-n \cdot t)\} \qquad (4.5)$$

For longer periods, one can assume a steady-state relationship and disregard the exponential term. This relationship is demonstrated in Figure 4.3, presenting measurements performed in a concrete house in which the uncontrolled airflows were minimized.

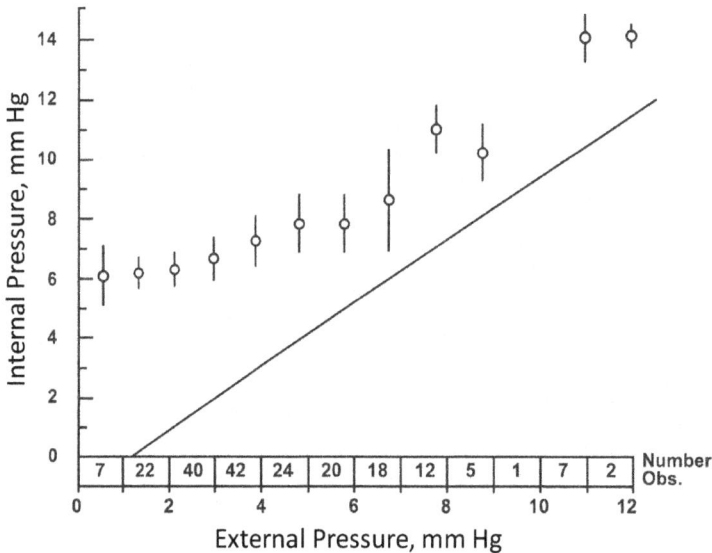

**Figure 4.3:** External and internal partial pressures of water vapor in a concrete house.

The line drawn at the 45° angle implies no moisture production in the house. The measured values show that the moisture production, related to the house ventilation $\{G/n·V\}$, is about $2.4 \times 10^{-3}$ kg/m$^3$. Measurements in office buildings show that $2 \times 10^{-3}$ kg/m$^3$ to

$3 \times 10^{-3}$ kg/m$^3$ can be used for calculations. Table 4.1, below, shows the effect of changing {G/n·V}.

**Table 4.1:** Relative humidity indoors (F) with T = 20 °C and different ratios of moisture production (prod) to ventilation (vent) for the selected outdoor climate conditions.

| OutdoorT °C, Φ %RH | Prod/ventilation 0 | Prod/vent 1 × 10$^{-3}$ | Prod/vent 2 × 10$^{-3}$ | Prod/vent 4 × 10$^{-3}$ |
|---|---|---|---|---|
| 5, 80 | 30 | 35 | 41 | 53 |
| 0, 85 | 22 | 28 | 34 | 45 |
| −5, 90 | 15 | 21 | 27 | 38 |
| −15, 90 | 6 | 12 | 18 | 29 |
| −25, 90 | 2 | 8 | 14 | 25 |

Dry air has a specific heat of 1 [kJ/(kg·K)], water vapor has a specific heat of 1.84 [kJ/(kg·K)], and the heat of evaporation of water is 2,500 [kJ/kg]; thus, condensation and evaporation processes involve a substantial quantity of heat.

## 4.2 Concepts related to moisture in porous materials

Several terms are used to describe the presence of water in porous materials. Water may be chemically bound, for example, in a gypsum crystal, or contained within a porous material. The latter is either adsorbed to the surface, absorbed by an existing water film, or condensed in the material capillaries (the latter is often called free water).

One often talks about hygroscopic water when it comes from moist air but has an equilibrium lower than 98% RH. For higher relative humidity values, we talk about over-hygroscopic or capillary water. Observe that this terminology is different from the terminology used in hydrology, where water has a positive water head, while a negative water head in building physics means that it is suspended inside a material. Furthermore, hydrology and hydraulics deal with one medium, namely a liquid. In building physics, we have a multicomponent mixture of liquid carrying dissolved salts, and air/vapor water mix. In the presence of a thermal gradient, in a transient condition, the following flows are involved:

(a) At a waterfront: a directional diffusion toward lower temperature, adsorption to the dry material surfaces in front and on the sides of water fingers, and water vapor condensation in micropores where the dew point is reduced by the pore surface curvature (i.e., pores with a diameter smaller than $10^{-7}$ m).

(b) At the developed water intrusion: flow of water adsorbed to the material surfaces, or water absorbed to the surface water, called "capillary liquid flow" (sometimes also called free water flow), water dispersion when the capillary water flow goes in a finger-like fashion, multidirectional vapor diffusion when evaporation takes place at the front of water fingers.

The connectivity of different pores includes pores with different shapes and sizes so that the capillary water flow will block a certain volume of air inside the liquid water field. The preferential filling of pores, namely the micropores and pores with temperatures below the dew point, entraps a fraction of air in the water field. It is, therefore, evident that a piece of material immersed in water will have a significant fraction of pores filled with air. This implies that the moisture content at capillary saturation is much lower than the open cell volume. It is also easy to realize that different ways of wetting, for example, under thermal gradient or by free water, will result in different moisture content at saturation. For this reason, we use two different terms: vacuum saturation and capillary saturation.

Vacuum saturation or maximum moisture saturation obtained under vacuum, is the total volume of open porosity accessible when air is evacuated using vacuum ($w_{max}$):

$$S = w / w_{max} \qquad (4.6)$$

Capillary saturation is defined as the end of a standardized free water intake test (see later text). This smaller moisture content, $w_{cap}$, will be reached with nearly maximum entrapment of air in the moisture field. The dimensionless, generalized variable used in moisture analysis, namely a degree of capillary saturation is a ratio between the actual moisture content (w) and the capillary moisture content

$$S_{cap} = w / w_{cap} \qquad (4.7)$$

where "w" is the actual moisture content, $w_{cap}$ is the maximum moisture content obtainable under the free water intake test. While eq. (4.7) is used in water content (moisture content) calculations, the presence of capillary hysteresis implies that for different hysteretic conditions, the $w_{cap}$ may be higher than the one defined by eq. (4.6).

The reader should remember that building physics (science) deals with predominantly transient processes. What we call moisture equilibrium is a statistical equality of two opposing processes of wetting (by condensation or water inflow) on one side and drying (by water outflow or evaporation at the water meniscus) on the other side. The reduction of moisture content in liquid phase we also call drainage.

Two concepts describe the quantity of moisture stored in a material: moisture content in percent of either weight or volume. To avoid confusion, we recommend expressing moisture content as the ratio of the mass of water to the dry volume of the material (kg/m³), or in percentage by volume, that is, m³ of water per m³ of the dry material. (To recalculate percentage by volume into kg/m³, multiply the value by 10.) We do not recommend using percentage by weight.

Compare clay-brick and polyurea foam, each with a moisture content of 10% by weight. To recalculate % by weight into % by volume, one multiplies it with the ratio of material density to water density, that is, 15% × 1.7 = 25.5%. The clay-brick with 1,750 kg/m³ density (with a solid matrix density of 2,700 kg/m³ and 35.2% porosity) means 263 kg of water per m³ or 26.3% by volume, a number close to the capillary saturation.

On the other hand, the polyurea foam, with a density of 9 kg/m$^3$ and a solid matrix of 1,050 kg/m$^3$ has 99.1% porosity. Its moisture content is 15% × 0.009 = 0.15% by volume. So, the same percentage by weight means an absolute saturation in one case or a minute fraction of pore volume in the other.

## 4.3 Moisture of the material in equilibrium with moist air: sorption isotherms

Porous materials exposed to a constant partial water vapor pressure in the air will reach an equilibrium moisture content characteristic for their fraction of micropores in the material (Figure 4.4).

Figure 4.4 shows sorption isotherms for a few common construction materials under wetting. Initially, water vapor entering the material is adsorbed to the material surface. With the increase in the relative humidity in the material's pores, the thickness of the adsorbed water film grows, modifying its density that approaches the density of free water. At this moment, the BET (initials of three authors of this theory) theory for multilayer sorption becomes valid, and capillary hysteresis begins. Further increase in moisture content is in the form of absorbed water and capillary water (the latter is likely to occur when water condensed in microcapillaries, Janz et al., 2001).

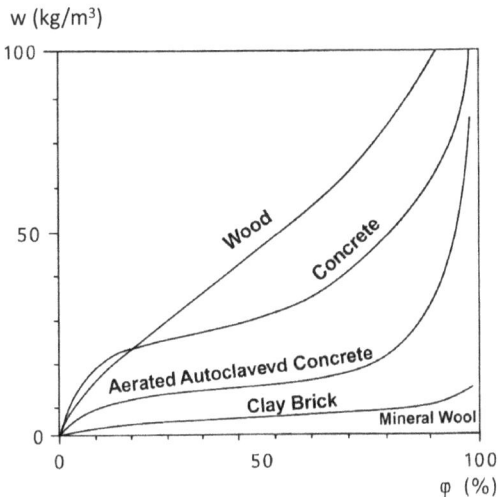

**Figure 4.4:** Typical sorption isotherms (wetting) for a few materials. In order of the growing hygroscopic moisture content: mineral wool, clay brick, aerated autoclaved concrete, concrete, and wood.

For multilayer sorption, the Thompson (Kelvin) law applies, and eq. (4.8), presented in Figure 4.5, shows the effect of pore size on the relative humidity above the pore water meniscus:

$$\Phi = \exp\{p_c/(\rho R_v T)\} \tag{4.8}$$

where $\Phi$ is the relative humidity, $p_c$ is the difference between pore water pressure and the atmospheric one [Pa], $\rho$ is the pore water density [kg/m$^3$], $R_v$ is the gas constant [J/(kg·K)], and T is the absolute temperature (K); when the wetting angle is zero, that is, in the presence of a significant layer of adsorbed water and disregarding the wetting angle (pre-wetted material), one may write:

$$p = 2\sigma/r \tag{4.9}$$

where are the surface tension of water (N/m) and r is the capillary radius (m)

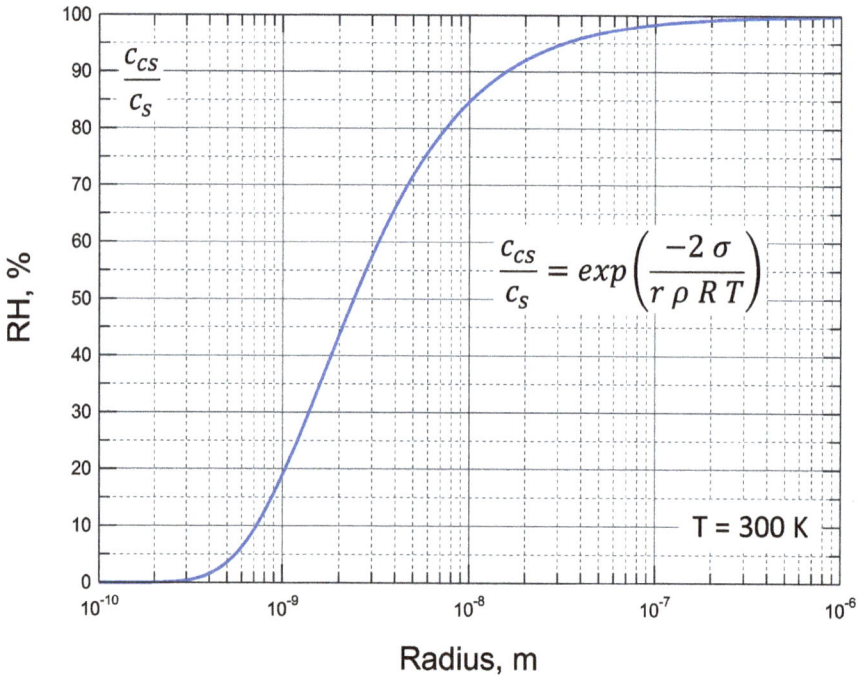

Figure 4.5: The saturation of water vapor in equilibrium with pore-water about the pore size.

## 4.4 Diffusion of water vapor through a porous material

The measurement of water vapor (WV) permeance (dry cup) is presented in Figure 4.6. WV diffuses from a conditioned environment (typically with 50% RH) to the desiccant (near 0% RH), and periodically measured weight increases of the desiccant cup give us the WV permeance value.

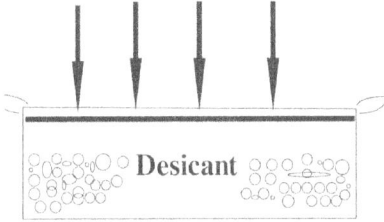

**Figure 4.6:** Dry cup measurement of WV permeance.

The isothermal value of WV permeance also represents the values obtained when diffusion occurs under a difference in temperature. As demonstrated by Hutcheon and Paxton (1958, HRef) and shown in Figure 4.7, the measured values and fully drawn lines of calculations based on water potential were determined under isothermal conditions.

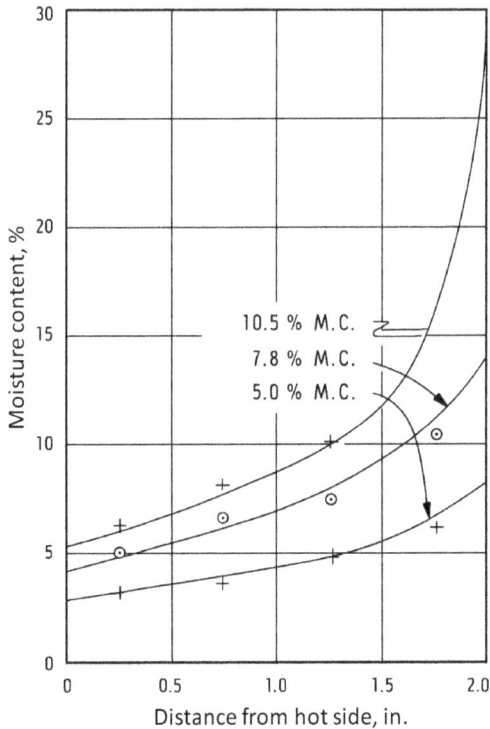

**Figure 4.7:** Moisture content in equilibrium is predictable from the sorption isotherm, even under thermal gradient conditions (Hutcheon and Paxton, 1958, HRef).

Figure 4.7 shows a good agreement between the values of moisture content measured under conditions of temperature difference with those calculated from isothermal tests (continuous lines).

## 4.5 Concept of capillarity

Figure 4.8 shows a wetting angle (angle formed in water between the solid surface and air-water interface) of (a) a drop of water on an active capillary surface ($\theta = 0$), (b) water on an inert surface, for example, glass ($\theta < 90$), and (c) liquid on a non-wetting surface (negative wetting angle), for example, mercury on glass ($\theta = 180$).

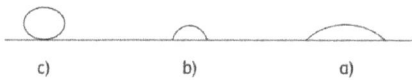

c)    b)    a)

**Figure 4.8:** Wetting angle in cases: (a) extreme wetting ability, (b) standard wetting, (c) non-wetting fluid.

While this picture represents the behavior of a liquid in contact with dry material, it may not be representative of moisture flow inside a porous medium because (as discussed later) the wetting by water vapor may take place ahead of the water movement and "pre-wet" the surface of the pores. The capillary rise of water in tubes with different diameters and inclinations is shown in Figures 4.9 and 4.10.

The equilibrium height that in this case represent water potential will be reached in the capillary tube at the height:

$$H = 2\sigma \cos\theta / (r\rho_w g) \tag{4.10}$$

where $\sigma$ is the surface tension of water [N/m], $\theta$ is the wetting angle, r is the radius of the capillary [m], $\rho_w$ is the density of pore water (a multilayer absorbed water has a density of water, in kg/(m$^3$), and g is the acceleration of gravity [m/s$^2$].

Equation (4.11) relates to Figures 4.9 and 4.10. The surface tension of water depends on temperature:

$$\sigma = 78 \ (1 - 0.0032\,T)\ 10^{-3} \tag{4.11}$$

where T is the temperature in °C.

## 4.6 Free water intake or rain on a material surface

Equation (4.12) presents an experimentally determined relationship

$$m = A\tau^{0.5} \tag{4.12}$$

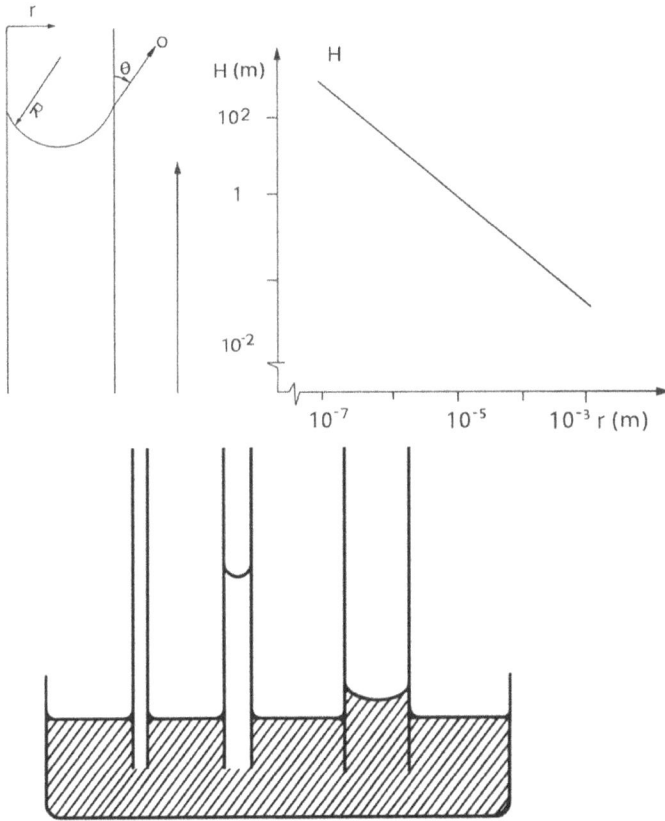

**Figure 4.9:** Capillary rise depends on the wetting angle, that is, the radius of the capillary.

where m is the mass of water absorbed from the beginning of the water intake process, A is an experimental constant called the water absorption coefficient, and $\tau$ is the test duration.

We present a few cases of A-coefficient testing to gain insight into factors affecting the water flow rate inside the porous material. Figure 4.11 shows that air pressure in front of the moving water will affect the water inflow rate. In contrast, Figure 4.12 shows that water supply interruptions have a lesser effect on the absorption coefficient. Of course, the flow with interruptions has a modified intercept. These experiments, as well as the work of Phillip (1957, HRef) on the theory of "sorptivity" in soil science, indicate that the assumption of the constant A-coefficient is limited to a short-term (few hours) experiment. Indeed, for most capillary materials in construction, this A-coefficient is not stable if the test is longer than 1–2 h (Plagge et al., 2004; Bomberg et al., 2005).

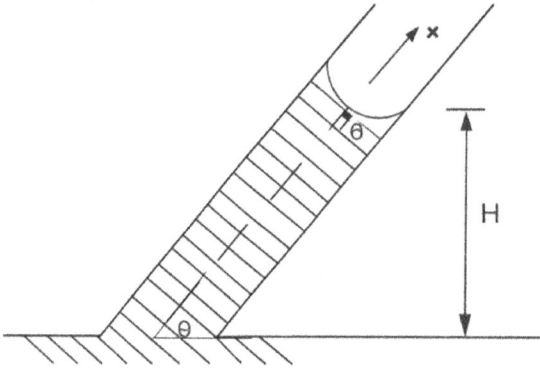

**Figure 4.10:** Water suction exhibited by a non-vertical capillary tube.

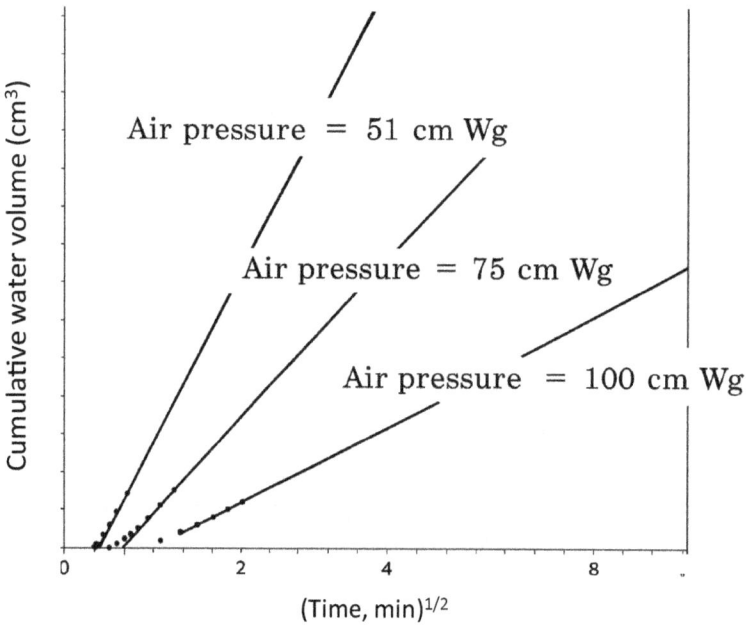

**Figure 4.11:** Cumulative infiltration as a function of (time)$^{0.5}$ for different air pressures ahead of the waterfront.

The next two figures show an apparent equilibrium after some time of free water intake. The first one shows it in a cumulative liquid inflow notation, and it com-

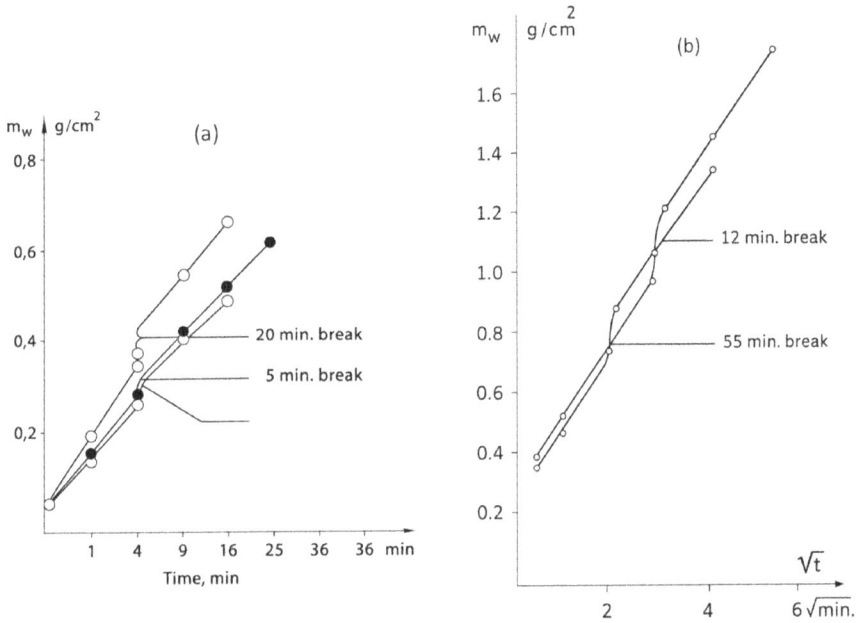

**Figure 4.12:** Cumulative infiltration volume a function of (time)[0.5] for tests with interruptions. Tests of Jansson (1965) reprinted from Bomberg (1974).

pares two series of tests performed in two different laboratories. TU Dresden used an automatic balance, while Syracuse University used a manual balance. Testing personnel in both cases were the same, and the results are the same (Bomberg et al., 2005).

Figure 4.14 uses a differential form of water influx to highlight that one cannot use the data from the initial testing period before the family of moisture distribution profiles is established, nor from the period when capillary saturation is approaching. Elsewhere (Bomberg et al., 2016), we show that the waterfront looks like several water fingers. The perpendicular dispersion between these fingers equalizes the field of water. This results in a transition visible in Figure 4.14 between 45 and 60 square roots of the second, to reach the capillary saturation at about 60 square roots of the second.

Finally, Figure 4.15 shows that in multilayered assemblies (e.g., brick and mortar or plaster), the capillary saturation value is quite different from tests performed on a single material. As capillary saturation is caused by air entrapment, the solubility and diffusion of air will slowly continue to increase the total measured water content. Air entry to the clay brick is prevented by the layer of plaster, and the experiment lasted for 3 years without significant change in the moisture content of the clay bricks.

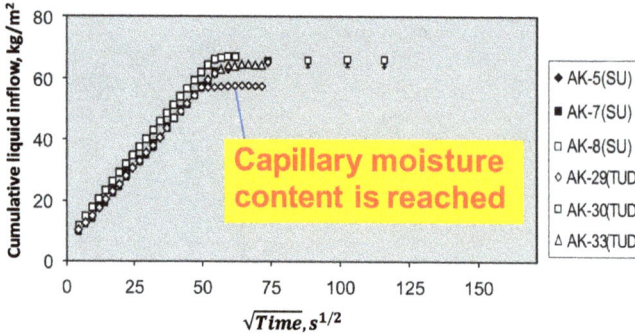

**Figure 4.13:** Cumulative liquid inflow in kg/m² measured on calcium silicate with automatic balance and manual weighing. Reaching the capillary moisture saturation is indicated in this figure.

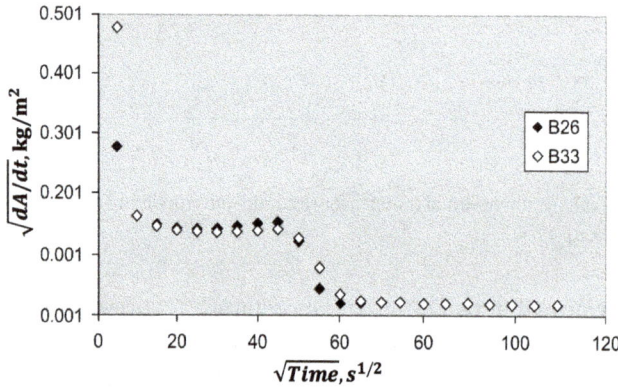

**Figure 4.14:** Differential measurement, dA/dt$^{1/2}$, displays the period with a constant A-coefficient.

The experiment shown in Figure 4.15 explains why some masonry from the Middle Ages could survive for ages, as long as the interior was heated in the winter and clay bricks could dry in cold seasons.

## 4.7 Moisture retention curves

The difference between pore water pressure and atmospheric pressure is called capillary suction ($p_c$) and represent thermodynamic potential of water, has already been defined in eq. (4.8), but it can be presented in a different form in the following equation:

$$p_c = \rho_w \, RT \ln \Phi \tag{4.13}$$

where $\rho_w$ is the density of water, $R_w$ is the gas constant for water, T is the absolute temperature, and ln $\Phi$ is the natural logarithm of the relative humidity of the pore air.

Observe that the gradient of temperature also affects the value of capillary suction (see below), causing the flow of water toward the lower temperature.

---

**Example 3.4:** A gradient of temperature creates the gradient of capillary suction
Example: $\Phi = 99\%$ ; $\ln \Phi = 1E - 02$
$\quad\quad \Phi = 99.9\%$ ; $\ln \Phi = 1E - 03$
$\quad\quad$ T1>T2 than $p_{c1} > p_{c2}$
$\quad\quad$ Moisture flows to lower T

---

Since sorption isotherm uses relative humidity, $\Phi$, and capillary suction uses a logarithm of the relative humidity in the pore air, one may consider the relation between moisture content and the logarithm of relative humidity for both capillary and hygroscopic fields together. We talk about the moisture retention curves (MRCs) in Figures 4.16 and 4.17. These two figures represent capillary forces acting within porous materials, and one can observe the large pressure range in these curves.

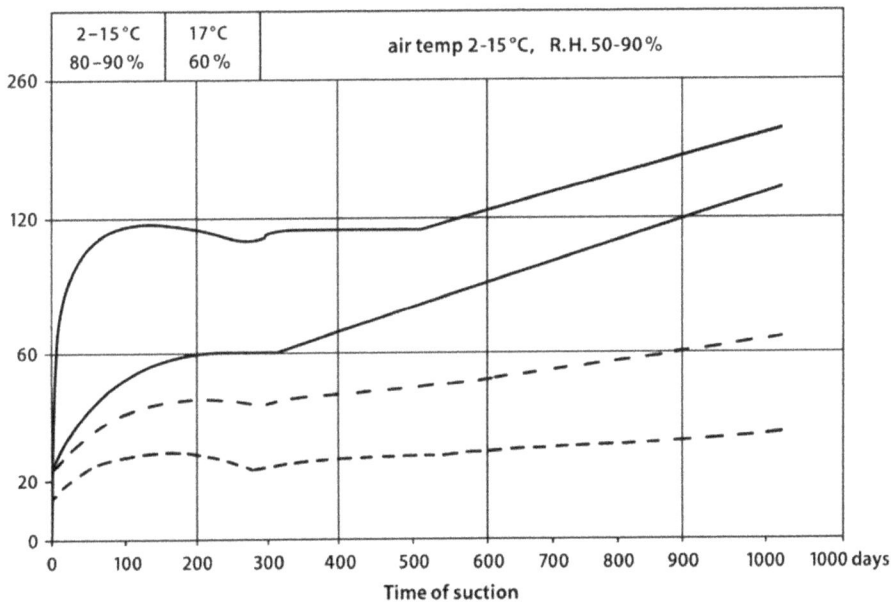

**Figure 4.15:** Capillary moisture content vs time measured on a combination of two types of clay bricks with two kinds of renderings. Removal of air from the pore-water system increases the capillary saturation – courtesy H. Kuenzel at IBP, Holzkirchen (HRef).

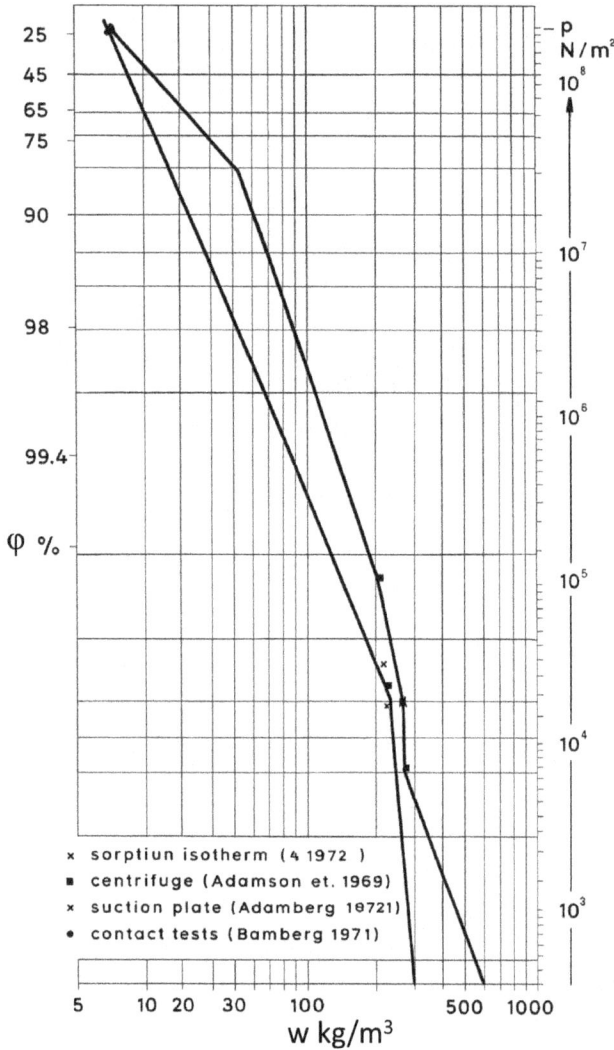

**Figure 4.16:** Moisture retention curves for aerated autoclaved concrete with a density of about 500 kg/m³ (Bomberg, 1974).

The hydraulic water flow with the open-ended capillary tube could lift a water column only to a 10-m height, that is, one-atmosphere pressure. The MRC (moisture retention curve) shown here goes far beyond 10 km of negative water head because those are the water pressures involved in moisture fixation to the micropores and microcapillaries. It is no surprise that the combination of microcapillaries and osmotic

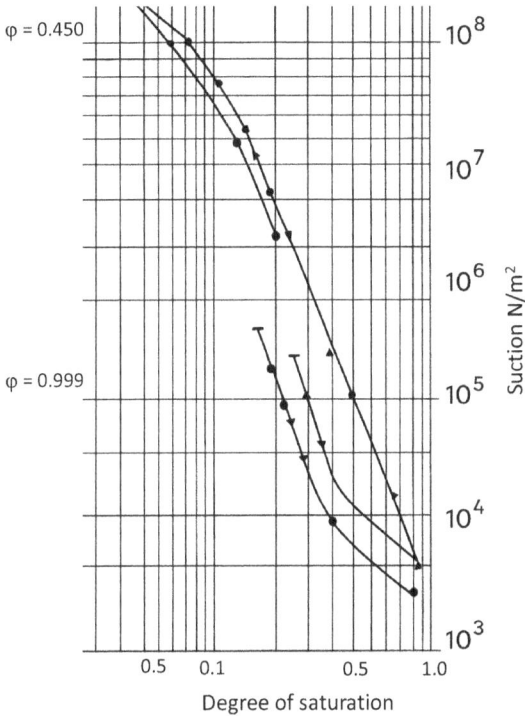

**Figure 4.17:** Moisture retention curves for wood fiberboard with a density of about 600 kg/m³ repeated with distilled water to show the effect of salinity changes.

pressures that takes place in nature can be seen in some tall trees, lifting water to several times the limit of one atmosphere.

Figure 4.16 shows tests performed with four different test methods by three researchers. Moisture content, determined to be in equilibrium during either the wetting or drying from absolute saturation, is presented against two variables: relative humidity (on the left side) and capillary suction (on the right side).

To reduce the effect of material variability (see later discussion), Figure 4.17 uses the degree of capillary saturation instead of moisture content. The test was performed using water saturated with wood fiber extract, and a maximum moisture content achieved under vacuum was used to start the drying run. Then, the following wetting run was interrupted, and when the subsequent wetting run was started using distilled water. The moisture retention curve did not follow the same trace. The shift in the equilibrium moisture content was measurable.

There are two reasons for such a shift:
1) A change in the solid phase organization, for example, pore size distribution, can be affected by material swelling and shrinking in the subsequent wetting and drying cycles (this often happens with organic, fibrous materials).
2) A change in the salinity of pore water, for example, caused by immersing the specimen in water with salinity different from that of the pore-water (instead of tap water, we place small pieces of the material and cook them to dissolve the salts and to obtain water with adequate salinity for precise measurements of material characteristics).

Figure 4.16 also indicates a region where capillary saturation is probable. Looking at the two regions shown on the wetting curve, one may assume that a moisture content of about 300 kg/m$^3$ corresponds to the capillary saturation point obtained under a prolonged wetting process. A slow drying from the vacuum-saturated specimen characterizes an almost identical moisture content. Thus, one may assume that both curves show the range of capillary saturation.

Figure 4.18 illustrates the effect of salinity on the pore water's energy, resulting in changes in the equilibrium moisture content. For sodium chloride (kitchen salt), the shift takes place at 75% RH, which corresponds to the suction value of $4 \times 10^7$ Pa, that is, at the end of the presented results.

## 4.8 Construction (built-in) moisture

Table 4.3 shows the typical amount of moisture introduced during the construction process.

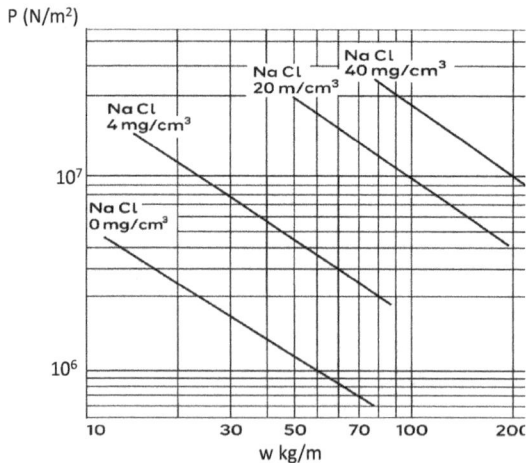

**Figure 4.18:** Effect of adding sodium chloride (kitchen salt) on the moisture retention of the clay brick.

**Table 4.3:** Built-in moisture in kg/m$^3$, chemically bound water, and amount to dry out.

| Material | MC as built | Chem. bound | Equilibrium at 50% RH | Amount to dry out |
|---|---|---|---|---|
| Concrete | 180 | 70 | 30 | 80 |
| Lime-cement mortar | 300 | 20 | 10 | 270 |
| Masonry | 80 | – | 10 | 70 |

## 4.9 Interactions between water liquid and vapor phases

Figure 4.19 shows inter-connected containers with moist air, each having the same total air pressure but a different fraction of water vapor. Air and water vapor diffuse in opposite directions.

Figure 4.19 shows the water vapor molecules diffuse from the right to the left tank, and air diffuses from the left to the right tank. The rate of each transport depends on the mean free path of the gas and the pressure difference. The linear gradient of water vapor concentration characterizes the flow shown in Figure 4.19.

The following equation defines the rate of water vapor diffusion:

$$Q = D \operatorname{grad} c \qquad (4.16)$$

where $D = (22.2 + 0.14\,T) \times 10^{-6}$ and temperature T is measured in °C.

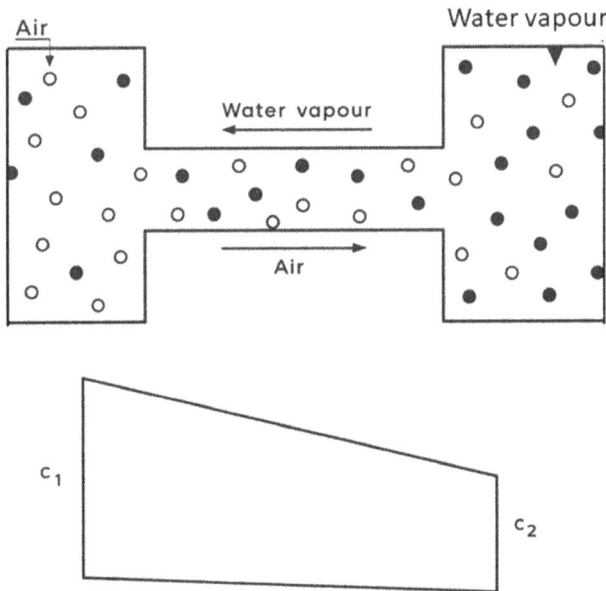

**Figure 4.19:** Diffusion of water vapor in the air.

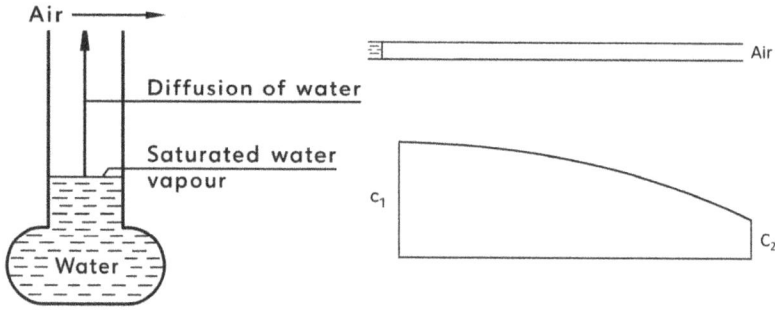

**Figure 4.20:** Evaporation and diffusion of water vapor through an air layer.

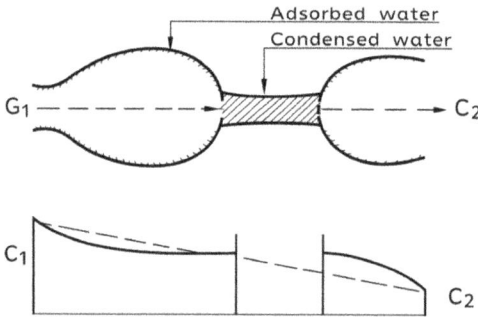

**Figure 4.21:** A general pattern of vapor concentration profiles in a porous material.

In contrast to this process, Figure 4.20 shows water evaporation from a capillary and subsequent diffusion through an air layer.

Figure 4.20 shows that the boundary conditions are constant. Yet, the concentration profile as a function of the distance from the evaporation surface is different because the potential of the pore water is described by eq. (4.11).[1] While the potential gradient is constant, it is the gradient of the natural logarithm of the relative humidity instead of the vapor concentration.

Figure 4.21 shows the diffusion of water vapor through an air layer and condensation, with subsequent evaporation on the other side of the capillary, as an illustration of the serial mechanism of water vapor transmission (WVT) in the porous body. Theoretically, the vapor diffusion rate should decrease with growing moisture content, since the volume of air available for vapor diffusion decreases with higher moisture

---

1 In thermodynamics, the potential of water in the porous material is defined by the difference between the Gibbs free energy in relation to the reference level of the free water table, that is, $\Delta f = RT \ln \Phi$. Observe that $p_c = \rho_w \Delta f$, and with $\rho_w$ (density of pore water) being constant and equal to the water density, the capillary suction may be used as a good descriptor of the water thermodynamic potential.

content. However, for hygroscopic materials, the WVT increases with relative humidity because of the contribution from the liquid phase. This is an apparent transport characteristic involving a multiphase system. For the sake of hygrothermal modeling, we need to establish the onset of the liquid phase transport function.

## 4.10 Movement of moisture causes heat transfer

Below, we demonstrate that phase change occurs during an isothermal water inflow into a porous material. The measurements were performed at the Lund Institute of Technology in 1970. They involved a container with water and the reference gypsum specimen, conditioned at a constant temperature for an extended period before the test. The test (Figure 4.22) was performed to demonstrate the interaction between heat and moisture flows during the "isothermal" environmental conditions. The temperature reduction shown in front of the water inflow to a porous material can only be explained by the cooling of the material during the evaporation of water.

Two trends in temperature changes can be observed in Figure 4.22. One starts soon after establishing contact between the specimen and water. For the next 4 h, the temperature of the whole specimen appears to be slowly decreasing. Superimposed on this profile are the results of each separate measuring point. The temperature profile recorded at the first measuring point shows a rapid decrease, reaching the limit of the scale within about 30 min. This period is longer for more distant points; for example, at 40 mm, the period is about 45 min, and at a 60 mm distance, about 1 h and 15 min. Generally, the temperature profiles look like moisture content profiles during the free water intake.

The explanation is easy when remembering that there is a mixture of air and water vapor at the waterfront; the more vapor, the longer the flow period. The amount of evaporation is smaller at the second measuring point and even less at the third measuring point. Even though the waterfront moves faster than the vapor can diffuse, the air displaced by water moves at the same speed as the water, and the superposition of vapor movement in the air is seen in Figure 4.22. The water/vapor ratio change at the waterfront results in the change in latent heat that is measured in this test. This is why the temperature profiles resemble moisture profiles observed by so many people before.

Nevertheless, the process depicted here demonstrated that one cannot separate liquid and vapor transports, even under so-called "isothermal" conditions. In effect, any calculation of moisture flow, one must include the thermal field and vice versa. In other words, one must perform hygrothermal (heat and moisture) calculations.

Another effect inferred from Figure 4.22 relates to the changes in the wetting angle of water. In the initial stage of water inflow into a dry material, one may talk about a specific wetting angle. However, after some time, the pore surface is wetted from water vapor diffusion ahead of the waterfront. The wetting angle under wetting and drying may be the same.

In the presence of a thermal gradient, vapor diffusing through the polymeric membrane will evaporate, diffuse through the cell gas, and condense at the next pore surface. Effectively, heat and moisture transfer cannot be separated. Even though our experiment dealt with an open-pore system, a similar situation is expected in closed-cell plastics since water vapor can penetrate through closed-cell walls.

It has been shown that vapor, which condenses in the middle of a porous material, evaporates and continues to flow toward the cooler temperature. It is safe to assume that Figure 4.22 shows multiple cycles of evaporation and diffusion of water vapor through the air, followed by subsequent condensation. These cycles constitute the actual pattern of moisture transmission in the porous body. It accumulates to a significant degree only when one of the following conditions is fulfilled:

1) Change in the equilibrium conditions introduces a phase change, reducing the vapor mobility, for example, temperature for a phase change transition.
2) There is a significant increase of resistance to vapor diffusion, causing the reduction of the flow rate, that is, the flow into the analyzed volume is much higher than the flow out of the book.

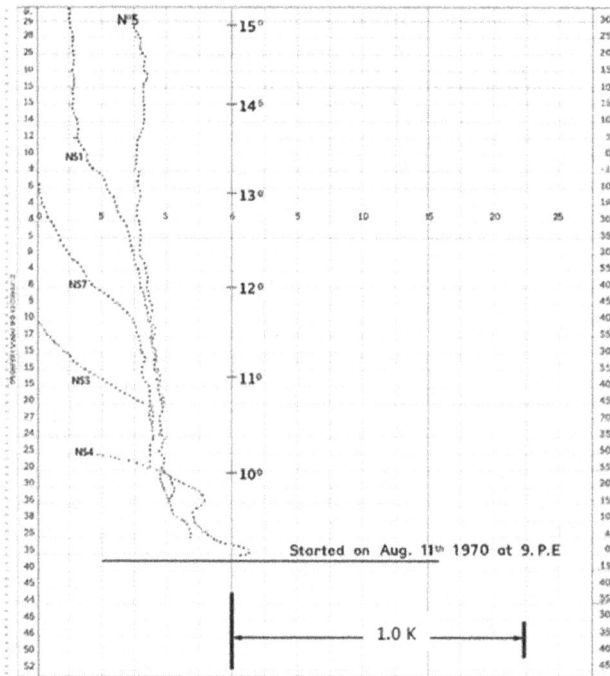

**Figure 4.22:** Temperature measured with 12 thermistors placed each 20 mm apart from the level of free-water intake into a gypsum column with insulated sides (from Bomberg, 1974).

This may happen when:
-   There is a reduction in the permeability of the next layer or interface of the material.
-   There is a significant reduction in vapor permeability caused by temperature change.

In a porous material, when the moisture content is above a specific range (typically when the equilibrium moisture content is higher than 50–60%), both the vapor and liquid-phases exist in the moisture flow. Figure 4.22 showed that one could measure a temperature difference caused by the evaporation–condensation cycle, even though conditions surrounding the specimen in which moisture flow takes place are strictly isothermal.

## 4.11 Thermally driven movement of moisture

Coupling between heat and moisture transport is one of the critical issues in hygro-thermal design because temperature gradients will always exist in building enclo-sures. Therefore, we present a whole experiment sequence highlighting the thermally driven moisture movement. Figure 4.23 shows specimen preparation to enclose a specified amount of water in the specimen, which is then placed in a heat flow meter apparatus (Figure 4.24) to study how water moves from the hot plate to the cold one. These tests were performed by Langlais (1983, HRef) at CRIR, Rantigny, France, and the results are shown in Figure 4.24a. In Figure 4.24a, one can observe a plateau of almost constant heat flux on one specimen from 3 to 8 h of the test. Kumaran (1987, HRef) used these data to calculate the water vapor permeability of the mineral fiber specimen and found (Figure 3.23) a good agreement.

**Figure 4.23:** (a) Moisture is sprayed on a surface and (b) placing the wet specimen in a sealed bag.

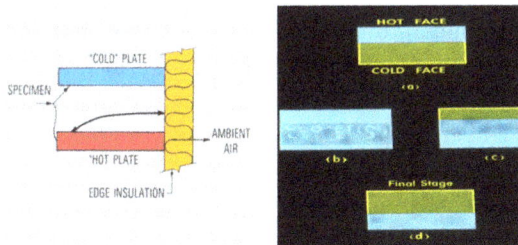

**Figure 4.24:** (a) Specimen placed in the heat flow meter apparatus with moisture on the hot side, and (b) moisture moves to the cold side.

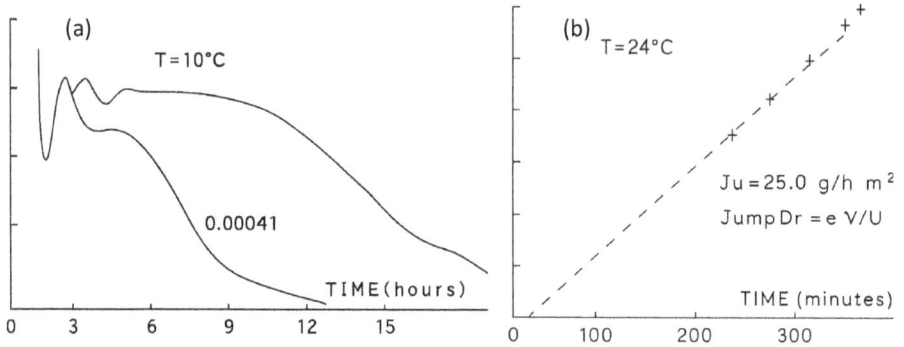

**Figure 4.25:** (a) Measured (apparent) thermal conductivity of the moist specimen (Langlais,1983) and water vapor flux calculated as a slope of moisture mass carried during the constant thermal conductivity "quasi-steady state" period of these measurements (Kumaran, 1987, HRef). (b) Measured from water vapor diffusion and calculated from thermal measurements (Figure 4.25a) showed that water vapor permeance differed by only 8%, indicating that the assumption of all water being evaporated and transferred as vapor until it condensed on the cold side of the specimen was entirely acceptable.

## 4.12 Material characteristics for moisture flow in hygrothermal modeling

Moisture flow rate (flux), q, is proportional to moisture conductivity, k, and the negative gradient of chemical potential, as previously discussed in eq. (4.13), while mass balance in 1-D notation is $dw/dt = dq/dx$. Moisture conductivity is not constant (Carmeliet and Roles 2001, 2002) but is highly variable and has a different thermal sensitivity for the vapor and liquid phases, and one must write the equation system for each of the phases separately.

As measurements relate to both phases together (Roles et al., 2003) one must measure several points above 90% RH to be able to establish the beginning of liquid phase contribution to the vapor phase transport. There are two German computer models in public domain that use different methods for including two effects of hysteresis.

The first effect relates to air entrapment. Unless there is a separate equation for air flow in the porous material (e.g., in soil science), the hygrothermal modeling is not valid above the capillary saturation. Therefore, using dimensionless moisture content – the actual above-hygroscopic moisture content, in relation to that at which the drying loop was started, allows an assumption of the whole volume of pores being used for the flow and approximates well the drying loop (Bomberg, 1974).

The second effect is capillary hysteresis, or more specifically, an effect of preferential filling of pores depending on the manner of moisture delivery. In this case, both computer codes, Delphin of TUD (Grunwald et al., 2003) and WUFI of IBP Fraun-hoffer bring us to the drying outermost curve. Delphin uses the moisture storage fac-tor, which is different for wetting and drying, while WUFI uses a variable WVT coeffi-cient that includes both liquid and vapor changes. In effect, if one calculates long-term changes, for example, a year, both codes are fine; for precise calculations of shorter transient changes, however, the Delphin is more reliable.

Typical hygrothermal models are parametric, that is, they are used to compare the effects of specific changes in input data, but they are not suitable for the real-time calculations (Bomberg et al., 2002). To calibrate (or validate) a parametric model, one must compare the calculation with a measured result in a special test (Bomberg et al., 2006). Some people talk about the variability of the hygric characteristics and believe that the calibration is used for ensuring a proper input. Whatever the reason is, the model calibration is now a standard procedure. Finally, a similar situation exists with the building system, and therefore we recommend a MAPE approach (monitoring and performance evaluation) to compare the model calculations with a specific field-measured data.

## 4.13  Control of rain penetration

Different aspects of controlling rain penetration are illustrated in Figures 4.26 and 4.27.

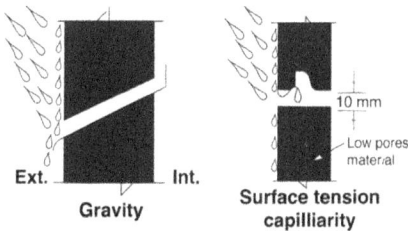

**Figure 4.26:** Control of rain penetration: gravity and capillary forces.

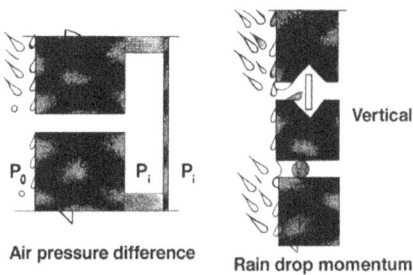

**Figure 4.27:** Elements of rain shedding – actual shedding, flashing with drips, and drainage. (Bomberg, 1974).

Figure 4.26 shows controlled rain penetration in the façade layer under the assumption that an air barrier eliminates any airflow through the wall assembly. The overlap in the water-shedding layer must be higher than the actual air pressure difference for the location. For drainage, we need a 3 to 4 mm air gap, but a smaller space may also be used for the gravity-induced evaporation-condensation sequence.

Figure 4.27 shows two ways of rain drop momentum control, typically used in large panel junctions. The caulking on the backer rod requires, however, periodic check if adhesive failure of the caulking compound does not introduce hair line cracks that may provide capillary suction.

The moisture management concept for insulated exterior walls in a cold climate, introduced by the Swedish researcher Johansson (1946, HRef) called for:
- An outer rain screen
- A drained and vented cavity
- Thermal insulation "between the actual wall and the rain screen"

The role of an air gap is illustrated in Figure 4.27, which shows contact experiments where dry material (aerated autoclaved concrete) was placed in contact with wet reference material (vacuum-cast gypsum) and sealed. Two experiments with different moisture contents of the reference material were conducted. Materials are in direct contact (curves 1), or moisture diffuses through a 1.5 mm air gap (curves 2).

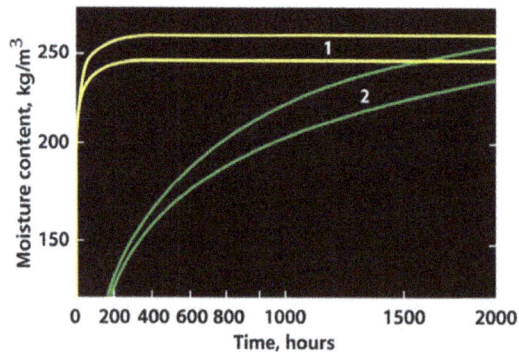

**Figure 4.28:** Contact experiments between a sealed pair of specimens, wet reference material, and dry test material: (1) in direct contact; (2) through a 1.5 mm air gap (Bomberg, 1974).

The moisture management concept for insulated exterior walls in a cold climate, introduced by the Swedish researcher Johansson (1946, HRef), called for:
- An outer rain screen
- A drained and vented cavity
- Thermal insulation "between the actual wall and the rain screen"

It is interesting to see how this concept evolved over time. The pressure equalization on the rain screen received much research, but it never became a practical solution because of the variable nature of the pressure fields. Yet, the concept of separation between the outer rain, radiative shield to provide some reduction of rain penetration and convective cooling in summer, or place the solar collectors is interesting. The thermal performance of this layer is not considered, and the exterior insulation would be placed on the inner part of the wall.

For wood or steel frame walls, we now consider a minimum of Rsi 1 in the continuous exterior thermal insulation (North American building codes) as a measure to reduce the effect of thermal bridges and eliminate water vapor condensation in frame walls, specifying as the minimum value to be used on the exterior of walls.

Yet, one may observe that people retrofitting existing buildings are now introducing an interior air gap to provide a possibility of drying old masonry construction and to create a low-energy assembly in buildings that otherwise would not be possible to upgrade.

## References

Bomberg, M. Moisture flow through porous building materials. Ph.D. Thesis at the Lund Institute of Technology, Report 52, 1974, LTH, Lund, pp. 188.

Bomberg, M.; Carmeliet, J.; Grunewald, J.; Holm, A.; Karagiozis, A.; Kuenzel, H.; Roels, S. *Position paper on material characterization and HAM model benchmarking*, Nordic Building Physics Symposium 2002, 2002.

Bomberg, M.; Pazera, M.; Plagge, R. Analysis of selected water absorption coefficient measurements. *J Thermal Envelope Build. Sci.* 2005, *28*(Jan), 237–245.

Bomberg, M.P.; Zhang, J.; Grunewald, J. *On validation of hygric characteristics used in heat, air and moisture models*. In 3$^{rd}$ Int Conf Bldg Physics, Aug 27–31, 2006, Montreal, QC, 2006.

Carmeliet, J.; Roels, S. *Determination of the Isothermal Moisture Transport Properties of Porous building Materials*. *J Thermal Envelope Build. Sci.* 2001, *24*, 183–210.

Carmeliet, J.; Roels, S. *Determination of the Moisture Capacity of Porous Building Materials*. *J Thermal Envelope Build. Sci.* 2002, *25*(3), January. 209–238.

Grunewald, J.; Haupl, P.; Bomberg, M. *Towards and engineering model of material characteristics for input to HAM transfer simulations – part 1- an approach, Introduction to material characterization for input to HAM-*. *J Thermal Envelope Build. Sci.* 2003, *26*, 343–366.

Janz, M.; Johannesson, B.F. Measurement of the moisture storage capacity using sorption balance and pressure extractors. *J Thermal Envelope Build. Sci.* 2001, 24(4), 316–334.

Lopez-Carreon, I.; Jahan, E.; Yari, M.H.; Esmizadeh, E.; Riahinezhad, M.; Lacasse, M.; Xiao, Z.; Dragomirescu. E. Moisture Ingress in Building Envelope Materials: (II) Transport Mechanisms and Practical Mitigation Approaches. Buildings. 2025, 15, 762. doi: https://doi.org/10.3390/buildings15050762.

Plagge, R.; Grunewald, J.; Häupl, P.; Bomberg, M. Analysis of water uptake experiments for building materials: methods, functions and parameters. In CIB W40 conference, Glasgow, 1–3 September, 2004.

Roels, S.; Carmeliet, J.; Hens, H. HAMSTAD, WP1: Final report Moisture transfer properties and materials characterisation, February 2003, K.U. Leuven, Belgium, 2003.

# Chapter 5
# Multidirectional heat flows

Heat, air, and moisture flows are inseparable, and we deal with them collectively under environmental control.

## 5.1 Mechanisms of heat flow

Basic equations for steady-state transfer of heat, electricity, water, or vapor through the material are identical (Figure 5.1, Bomberg et al., 2016). The potential that governs these flows: that is, absolute temperature, electro-motoric force or voltage, water head, and concentration of water vapor, each in relation to the resistance of the flow, give us the density rate of the flow (the flux). Thus, the energy flux is directly proportional to the temperature difference and inversely proportional to the thermal resistance of the material layer. In turn, the resistance is directly proportional to the thickness of the material layer and inversely proportional to the material layer's thermal conductivity (permeability).

Δ potential: water flow  ΔH (water head)
electricity  Δ U (EMF, voltage)
heat        Δ T (temperature)
vapor       Δ p (partial pressure)

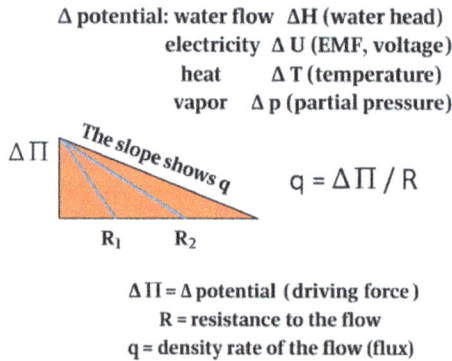

$$q = \Delta \Pi / R$$

Δ Π = Δ potential (driving force)
R = resistance to the flow
q = density rate of the flow (flux)

**Figure 5.1:** In the steady-state condition, the flow density rate is directly proportional to the potential (here: absolute temperature) and inversely proportional to the resistance (here: thermal resistance of the material layer).

For the volume element dv, (dv = dx·dy·dz), the heat flux is:

$$q = -\lambda \, \text{grad} \, T \tag{5.1}$$

where $\lambda$ is the thermal conductivity coefficient, the volumetric source or sink is "s," and during time "dt," the change of the storage term is expressed as $d(CT)/dt = (\rho c)dT/dt$. Putting it together, one may write that the change in storage of the element dxdydz over time dt is equal to the divergence of heat flux plus the work of the heat source or sink "s":

$$(\rho c) \, dT/ \, dt = \text{div} \, (-\lambda \, \text{grad} \, T) + s \tag{5.2}$$

https://doi.org/10.1515/9783112217023-005

## Concept of material storage

(mass or energy; here thermal energy)

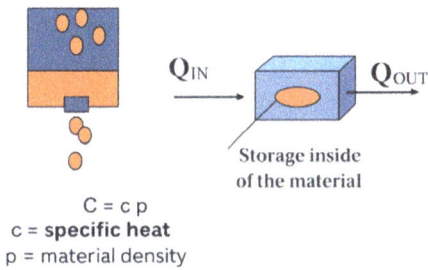

$C = c\, p$
c = **specific heat**
p = material density

**Figure 5.2:** In transient state conditions, the difference in material storage over a period (Δt) is equal to the difference between the incoming and outgoing heat fluxes (ΔQ).

In discussions of heat transfer, we use the following terminology: thermal conductivity coefficient, λ, W/(mK); thermal resistivity (inverse of λ), r, (m·K)/W; thermal resistance of the material layer, R, (m² K)/W, or thermal resistance of the building enclosure that is a sum of resistances of all material layers; thermal conductance of the layer (inverse of R), Λ, W/(m² K); thermal transmittance of the wall, U-value, that is an inverse of the sum of the film surface resistances and all material layers, W/(m²K); film surface coefficient, $h_s$, W/(m²K); or film surface resistance (inverse), $R_s$, (m²K)/W; specific heat, c (actually $c_p$), J/(kg·K); the heat capacity of material volume, C = ρc, J/(m³ K); surface emittance, ε (–); thermal diffusivity, D = λ/(ρ·c), m²/s; thermal effusivity, b = (λ·ρ·c)^½, J/(m²s^{1/2}K); and material density, ρ, kg/m³

There are three fundamental mechanisms of heat transfer: conduction (Figure 5.3), radiation (Figure 5.4), and convection (Figure 5.5).

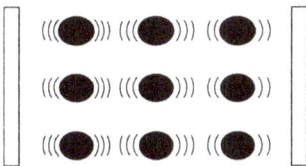

**Figure 5.3:** Vibration of molecules transfers heat by conduction in the solid material, from the end with a higher temperature to the end with a lower temperature.

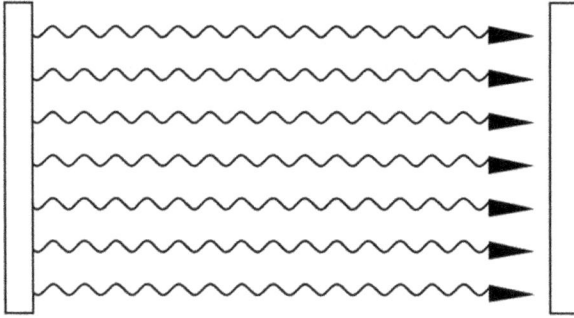

**Figure 5.4:** Electromagnetic radiation can transfer heat through air or vacuum.

The mass flux of gas with the density $\rho$ (kg/m$^3$) and the velocity v (m/s) is: $q_m = \rho$ v, kg/(m$^2$s), and the total energy carried by the gas at constant pressure becomes $q = c_p$ T $\rho$ v, J/(m$^2$s). Convective air flow can be either laminar or turbulent or both (Figure 5.5).

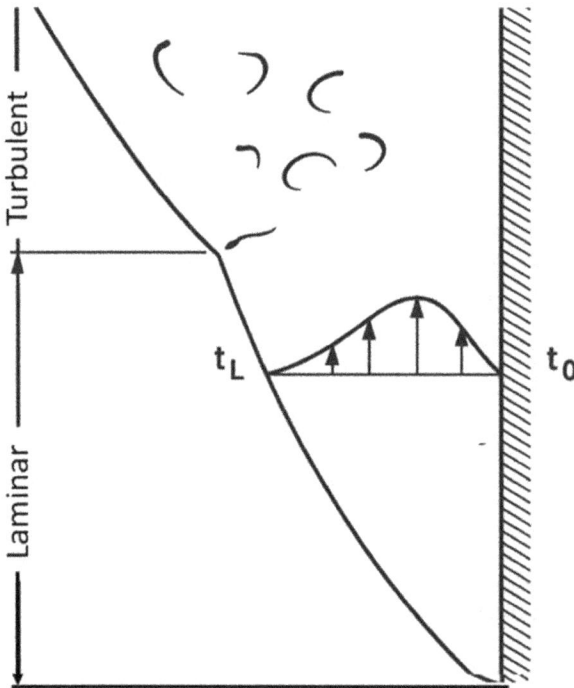

**Figure 5.5:** Two types of convective air movement near the vertical plane with a temperature higher than the fluid temperature. In the initial stage, one sees a laminar flow in the boundary layer, developing into a turbulent flow after some distance of the movement.

Heat transfer by convection may take place when convection is natural, that is, buoyancy forces cause air to move, or forced, when air pressure difference causes air movement next to the material with different surface temperatures.


## 5.2  Factors affecting multidimensional heat transfer

Let us consider heat flow through a wall in a cold climate during winter. In section A (through the center of the stud) and in section C (through the center of insulation), where flow is one-dimensional (1-D). But what is the flow pattern in section B, located near the thermal bridge?

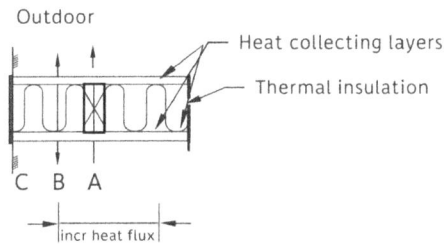

**Figure 5.6:** A thermal bridge created by a wood stud breaking a continuity of thermal insulation.

The wall shown in Figure 5.6 has an oriented strand board (OSB) on the exterior and gypsum (drywall) on the interior. In section B, the path of thermal energy goes first along the drywall (we call it a heat-collecting layer), coming to the stud, and then through the wood stud to the cold side. The lateral transport through the drywall and OSB (which, in this case, are also heat-collecting layers) will take place as long as the thermal resistance of the path B is smaller than that of section C. Increased heat flow causes temperature depression on the part of the wall surface that is denoted as the range of thermal bridge on the part of use wood has higher thermal conductivity than thermal insulation.


## 5.3  Effect of heat-collecting layers

Figures 5.7 and 5.8 expand our understanding of the role of heat-collecting layers. Figure 5.7 shows that the impact of a thermal bridge becomes stronger if a heat-collecting layer is conductive. Figure 5.8 shows a precast, large concrete panel with 50-mm (2 inch)-thick insulation and 50 mm interior and 100 mm (4 inch) exterior concrete panel. The joint between two panels constitutes a 60 mm (2 3/8 inch) break in the thermal insulation layer. We calculate the thermal resistance of the wall using two simplified methods. The first is called the "parallel flow method." With the input of the 1-D flows (sections A and C in Figure 5.6), this method calculates the area-weighted R-value. It means that each of these two areas is treated as if they had 1-D flow and there was no heat exchange between them.

We use the second simplified model, called the "isothermal planes method," which assumes each material layer conducts as much heat as its property allows laterally. Its surface has the same temperature in the cross section through the thermal bridge as it does through the insulation field. The parallel flow method gives the lowest value of the heat flow. It, therefore, overestimates the thermal resistance, while the isothermal planes method presents the other extreme case and underestimates the result. The actual result is always between these two cases (see later text).

Looking at Figure 5.8, we may ask two questions, namely:

– what imaginary width of thermal insulation break should be added to make the parallel heat flow model give correct results, and
– how is the position of insulation within the cross section of the wall affecting the results?

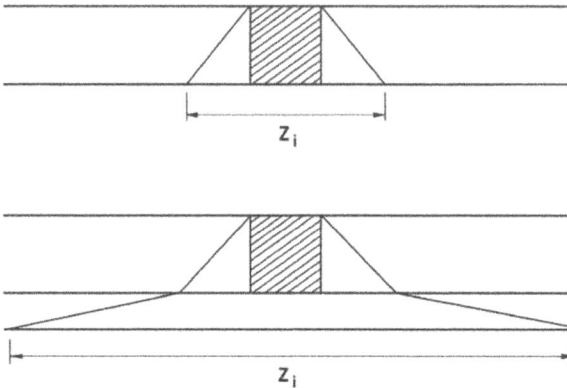

**Figure 5.7:** A range of the thermal bridge, $Z_i$, without and with heat-collecting layer.

To answer these two questions, the results of several finite difference calculations are shown in Figure 5.9, which generalizes the case shown in Figure 5.8. The middle curve in Figure 5.9 represents the combined thickness of both layers being 15 cm (6 inch). For $x = 0$, the insulation is placed on the interior side; for $x = 1$ (100% of the thickness), it is placed on the exterior.

A 60 mm (2 and 3/8 inch) wide gap in insulation on a 3m (10 ft) span represents only 2% of the area. Yet, Figure 5.9 shows an increase of 90 mm, bringing the correction for the thermal bridge effect to 5% of the area that the R-value of the thermal bridge must replace. Figure 5.9 also shows the effect of insulation placement. The correction is as significant as 120 mm (4 ¾ inch), that is, 6%, if the insulation was placed in the middle of the wall. In such a case, if one wanted to use a simple correction based on the surface areas to estimate the actual effect of the thermal bridge, one must use a 180 mm (7 ¼ inch) insulation break, not the 60 mm (2 3/8 inch) that was shown on the design drawings!

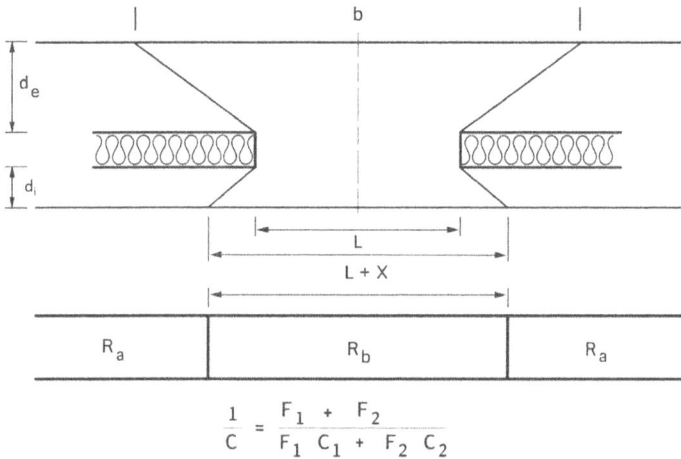

$$\frac{1}{C} = \frac{F_1 + F_2}{F_1\, C_1 + F_2\, C_2}$$

**Figure 5.8:** Effect of the placement of insulation within the wall.

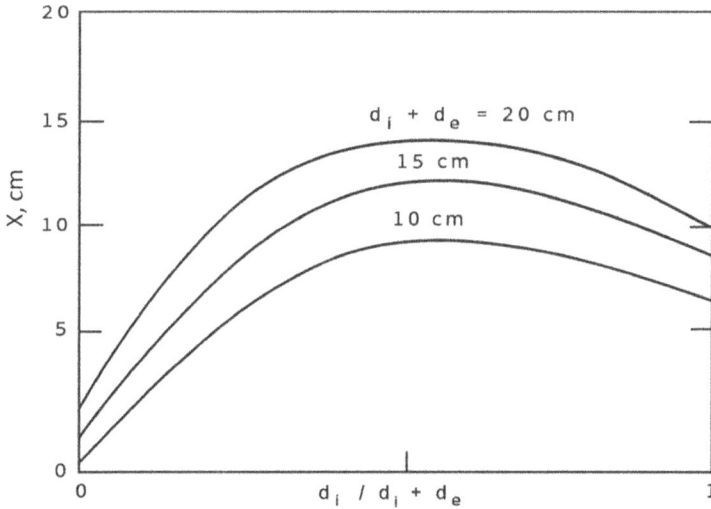

**Figure 5.9:** Distance X, to expand the width of the thermal bridge for the use of the parallel heat flow model (Bomberg et al., 2016).

Figure 5.9 shows that the same level of insulation has different effects on the thermal performance of the wall depending on the location of this insulation. This issue was also studied by Berthier (1960), who compared the placement of thermal insulation on the warm side (a) or cold side (b) of the width of the thermal break. The latter research is shown in Figure 5.10 in the 7.5 cm (3 inch) to 45 cm (18 inch) with two intermediate steps in between.

## 5.4 Effect of insulation placement

Results with insulation placed on the cold side of the wall (series b) show that the wall's internal surface temperature reduction is more significant when the width of the thermal bridge increases. Temperature depression was 7 °C for the 75-mm gap, but it rose to 9 °C at the 150-mm gap and 13 °C at wider gaps. Corresponding temperature depressions in series a), with the insulation on the warm side of the wall, are 10.2 °C for the first, 11 °C for the second gaps, and 13 °C for gaps 225 mm or larger. Evidently, with gaps of 225 mm (9 inch) or more, the placement of insulation did not make any difference. Yet, with smaller holes, which are more likely to occur in practice, the placement of insulation on the outside is preferred.

While thermal bridges in prefabricated concrete panels are expected to significantly impact the wall's thermal performance, the results of similar research per-

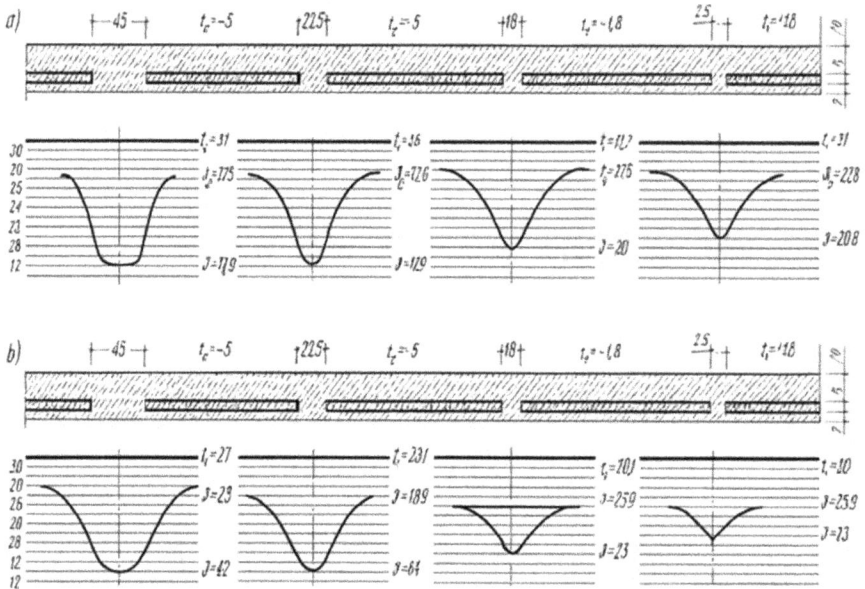

**Figure 5.10:** Effect of a break in polystyrene insulation on the temperatures of the concrete panel; test conditions: indoor 30 to 31 °C; outdoor −5 to −6 °C, one case 1.8 °C (from Berthier, 1960).

formed on wood construction by Ossowiecki (1964), as shown in Figure 4.11, are sur-
prising. The temperature on the insulation surface was 14.6 °C, and on thermal bridge
surfaces varied from 10 °C (50mm) to 11 °C (20 mm). Series b), where cement board
and drywall were applied, showed only a small change in the middle of the insulation
(0.2 °C). Yet, the presence of heat-collecting layers increased the thermal bridge temperatures from 0.6 to 1.1 °C.

Also, series c, where an insulation material was used instead of a gypsum board,
showed a significant increase in temperature on the thermal bridge surfaces. Replacing drywall with the same thickness of porous fiberboard increased temperatures
from 2 to 2.5 °C (see series c). Using a 25-mm-thick layer of the same fiberboard increased temperatures on the thermal bridge from 3.5 to 4 °C (Ossowiecki, 1964).

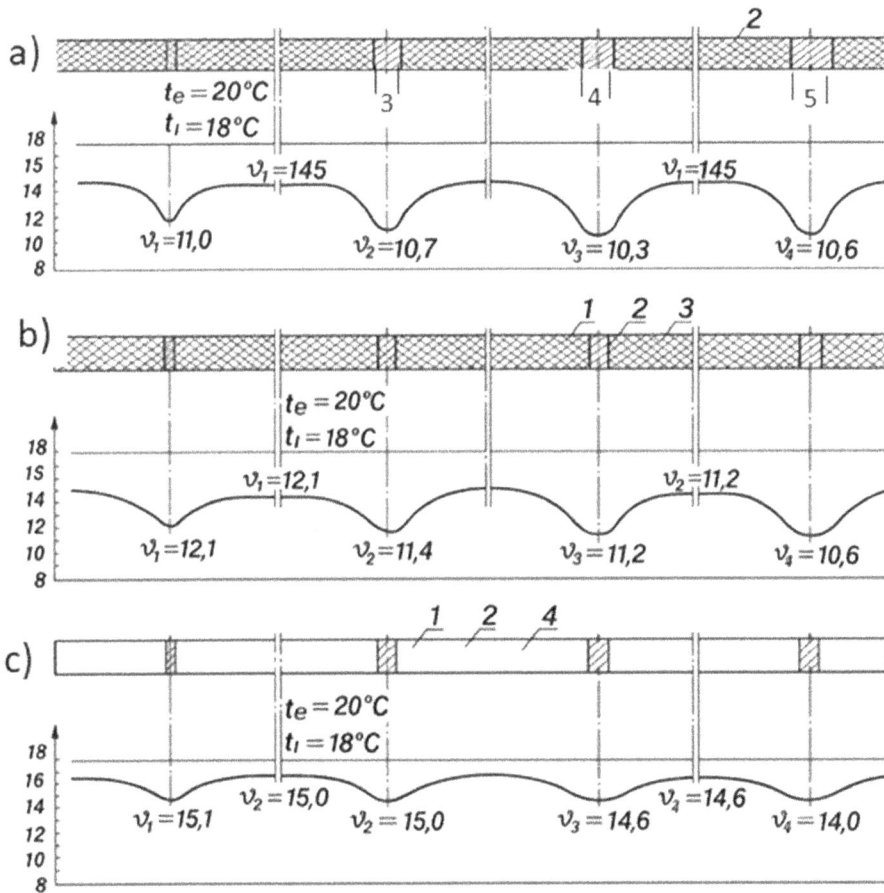

**Figure 5.11:** Temperature profiles measured on the wood-frame wall with 40 mm expanded polystyrene.
The heat-collecting layers are: (a) none, (b) outside cement board, inside drywall, (c) outside cement
board, and inside porous fiberboard (insulating sheet) – test conditions: 18 and −20 °C.

Figure 5.11 shows that conductive or insulating materials can be used in building enclosures to avoid condensation. The conductive layers will draw the thermal energy from elsewhere and deliver it to the point of thermal depression; the insulating layer will reduce the lateral transport and increase the thermal resistance of the wall. While the conductive layer may increase the overall heat losses, it can efficiently reduce the risk of surface condensation. This approach is often used in designing curtain walls and window frames. If a thermal break does not eliminate the risk of condensation, one may use a conductive layer on the warm side. By locating on the warm side a heat-collecting layer with a much larger area than on the cold side, one may increase the temperature on the inner side of the thermal bridge.

Indeed, these two aspects of the thermal bridge will always need to be reconciled. Figure 4.9, considering an increase in heat transfer, showed that placing insulation on the warm side is preferred. Figures 5.10 and 5.11, dealing with temperature depression caused by the thermal bridge, show that exterior placement of thermal insulation is desired.

## 5.5 Effects of boundary conditions and corners

Consideration must also be given to the interaction of radiation and convection at the surface of the thermal bridge (film resistance on the surface). An example of such interaction is shown in Figure 5.13. Even though the wall is homogeneous, heat transfer conditions in the corner are different. Firstly, the surface film coefficient is smaller because the view factor for radiation in the corner is smaller than on a clear wall. Secondly, the convective air currents in the corner are also smaller. Thirdly, the surface area on the exterior is more significant than on the wall's interior. The temperature on the interior surface measured from the corner to ½ m (1.5 ft) is shown in Figure 5.12. These measurements were performed under actual field conditions with an outdoor temperature of about −7 °C and an indoor temperature of 20 °C.

One may observe that the north wall (right side of the corner) had a slightly higher surface temperature (13.5 °C) than the west wall (12.8 °C). The corner temperature dropped to 8.5 °C, that is, a depression of 5° compared to the center of the north wall. (Note: The difference in thermal performance of walls with different orientations is also affected by differences in long-wave radiation, i. e., radiation between the wall surface and the environment.)

## 5.6 A concept of thermal insulation efficiency

A concept of thermal insulation efficiency uses a ratio between the actual and the nominal R-value, that is, the average calculated or measured thermal resistance of the assembly in relation to the nominal R-value. Typically, the nominal R-value is the sum

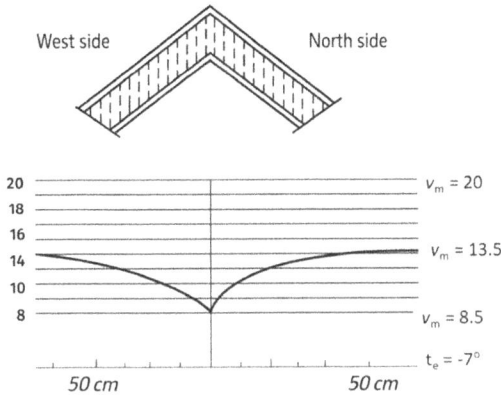

**Figure 5.12:** Temperature profiles measured on the interior of a homogeneous wall (Ossowiecki, 1964, from Bomberg et al., 2016).

of the thermal resistances of all assembly components without considering multidimensional flows. At this stage, we need to focus on the difference in efficiency between the interior and exterior placement of the insulation.

Compare a traditional 2 × 4 wood frame (OSB and gypsum board sheathings) filled with cellulose fiber insulation and a 2 × 6 frame wall with the same type of insulation. We use an advanced 2-D hygrothermal model and present the results in Table 5.1. In 2-D calculations, we determine the average heat flux over the range from the center of the stud to the center between studs and use it to calculate the mean R-value of the assembly. The ratio of the mean R-value to the nominal R-value is the efficiency factor for the insulation in the wall assembly. The difference between these two R-values, in relation to the nominal R-value, is the reduction of thermal resistance, typically presented in percent of the nominal R-value. The reduction of the R-value shown in Table 5.1 does not account for other thermal bridges in the wall, for example, at corners and windows, or the effects of moisture transport.

The effect of thermal bridges on residential construction is documented in the ORNL calculator: http://www.ornl.gov/sci/roofs±walls/AWT/InteractiveCalculators/rva lueinfo.htm that estimates an average reduction of R-value caused by all thermal bridges in 2 × 4 wood-frame walls (corners, windows, and studs) as 18%, and for 2 × 6 wood-frame walls as 28%.

**Table 5.1:** Percent reduction from nominal R-value caused by studs in the nominal "2 × 4" (40 × 90 mm) and "2 × 6" (40 × 130 mm) wood-frame walls with cellulose fiber insulation; metric values are placed in brackets.

| Cavity insulation thickness, mm | Resistance per m of thermal insulation (m K)/W | Thermal conduct. coefficient W/(m K) | Nominal R-value of the wall, the center of the cavity (m²K)/W | R-value from the 2-D calculation (m²K)/W | Reduction from nominal R-value, percent |
|---|---|---|---|---|---|
| 89 | 24.6 | 0.041 | 2.51 | 2.17 | 13.3 |
| 140 | | | 3.76 | 3.21 | 14.5 |

## 5.7 Estimating the effect of multidimensional heat flow

While accurate hourly climatic data for heat and moisture transfer must be included in the hygrothermal calculations, simplified thermal resistance calculations are sufficient in the conceptual design stage.

### 5.7.1 Serial and parallel flow models

A simplified equation that can be used for most wall assemblies (except for steel construction) is presented below:

$$R = (R_1 + m R_2) / (1 + m) \tag{5.1}$$

where $R_1$ is the thermal resistance calculated from the parallel path model, and $R_2$ is the thermal resistance calculated from the isothermal planes model.

These two models are discussed in the ASHRAE Handbook. The parallel path model gives too high a thermal resistance value, and the isothermal plane model, which assumes a perfect equalization of temperature within each parallel wall section, provides too low a value. Equation (5.1) states that the actual thermal resistance of the wall is between these two limits. A 1960s Russian thesis analyzed different insulated and hollow blocks, showing the value of m varying between 0.1 and 80, with the most probable value of m = 2. Therefore, if the difference between R1 and R2 is small, Polish standards used m = 2 (Plonski, 1965, HRef[1]). Brown and Schwartz (1987, HRef) showed that m = 1 approximated most insulated wood-frame walls. Garrett (1979, HRef) showed that the relation R2/R1 varied between 1 and 1.7 for lightweight slotted concrete at one end and dense concrete with foam inserts at the other. Shu et al. (1979, HRef), Valore (1980, HRef), and Valore

---

[1] Only selected references are given in this handbook; others are provided for the documentation of historic sources used.

et al. (1988, HRef) found that the isothermal planes model gives a much closer approximation to the measured R-values than the parallel path model (indicating a large m-factor).

The differences between R1 and R2 were much more significant for steel stud construction than those for masonry construction. The insight into heat flow patterns
through steel stud construction led to the development of a method that uses a network of resistances. Starting from textbooks on heat transfer, the Swedish Research
group (Johansson and Aberg, 1981, HRef; Johansson, 1981, HRef) developed a guide for
the design of highly insulated steel stud construction (Johansson and Andersson, 1989,
HRef). These examples were recently verified with 3-D computer models (Blomberg
and Claesson, 1993, HRef). Trethowen (1995, HRef) proposed an alternative approach
that incorporates contact resistance.

Recognizing these effects, Bomberg and Lstiburek (1998) proposed to use eq. (5.1)
with the following m-factors:

**m = 1.0,** for wood-frame construction and insulating masonry block

**m = 1.4,** for ceramic masonry blocks and steel stud construction, if the adjacent
layer has thermal resistivity higher than wood (insulation)

**m = 1.8,** for concrete masonry blocks and steel stud construction, if the thermal
resistivity of the adjacent layer is equal to or lower than that of wood

## 5.8  Effect of thermal bridges in wall assembly

We analyze the junction between the wall and floor in masonry (Table 5.2). Applying interior insulation made the situation worse (case 2b). This action reduced the TB temperature from 17.9 to 12.6 °C. The use of interior insulation eliminated the heat-collecting layer
on the interior surface (reducing the lateral heat flux to the thermal bridge). While the R-
value in the middle section of the wall was increased, the heat flux increased in the
bridge. Effectively, interior thermal insulation increased the effect of the thermal bridge.

Conversely, exterior insulation (case 2c) has a positive effect. Even though there
was no change in the temperature depression when external insulation was used, the
heat flux through the thermal bridge itself was reduced. The values of heat flux
shown in Table 5.2 are 0.17 W/m$^2$ for interior insulation and 0.005 W/m$^2$ for the exterior, that is, less than 3% of the previous case. Quick 1-D calculation for the data published in Table 5.2 reveals that the clay brick masonry had R = 0.83 (m$^2$K)/W and insulation R = 2.5 (m$^2$K)/W (e.g., 3-inch extruded polystyrene). The increase of R-value in
the actual wall was R$_o$ = 1.8 (m$^2$K)/W, giving the efficiency factor of 72%.

Case 2a, shown in Table 5.2, relates to the wood-frame wall with slightly higher
thermal efficiency than the insulated masonry wall. The same thermal bridge (concrete floor) now carries an extra heat flow of 0.16 W/m$^2$. Yet, the higher efficiency of
the wall means that this TB reduces overall thermal performance more than in the
previous case, that is, 36% (10/28) as opposed to the 27% in the previous case. Temperature reduction is now from 17.6 to 15.4 °C.

**Table 5.2:** Changes in U-value (inverse of R-value) and thermal depression caused by the thermal bridge created by a concrete floor in typical masonry constructions (Sandin, 1990, HRef).

| Drawing | Description | Additional flux (W/m²) | U-value $(W/(m^2 \cdot K))$ | | Temperature decreases due to the thermal bridge |
|---|---|---|---|---|---|
| | | | Excl thermal bridge | Incl thermal bridge | |
| | a) Floor joining wood-frame wall | 0.16 | 0.28 | 0.38 | 17.6 → 15.4 |
| | b) Interior thermal insulation of 1 ½ inch brick wall | 0.17 | 0.30 | 0.38 | 17.9 → 12.6 |
| | c) Exterior thermal insulation of 1 ½ inch brick wall | 0.005 | 0.30 | 0.30 | 17.9 → 17.9 |

To highlight the increased effect of thermal bridges for better-insulated walls, we present a series of 2-D calculations for wood-frame walls with different classes of cavity thermal insulation, both with and without exterior insulation. We use thermal insulation classes to avoid talking about specific products. It is necessary because many generic types of materials have significant differences in thermal performance (e.g., fiberglass or open-cell foam products).

There is another reason for the reduction of thermal performance of materials, namely sampling for the test. In the US, manufacturers of open-cell spray foam claim thermal resistivity in Imperial Units between 3.5 and 4.0 (ft² * h * °F/ BTU*in), while data measured (Table 5.3) on materials collected from field applications range from 3.12 to 3.56 ft² * h * °F/BTU*in) because the federal government allows samples to have a 90% limit.

While differences between specimens from the same batch were minor, all foams were near 90% of their claimed performance (the legal limit in the US).

Table 5.5 shows that the use of exterior insulation improves the overall efficiency factor of the insulation. This observation supports two conclusions:

1) Wood-frame walls in all climates should use a minimum of RSI 1 (R5.6) exterior insulation.
2) High-performance thermal insulation should not be used in cavities but rather on the exterior of the frame wall.

**Table 5.3:** Results of the thermal conductivity tests for foams coded A1, A2, and A3.

| Spec. code | Thickness (mm) | Thermal conductivity (W/m·K) | Thermal resistance (m2K/W) | Thermal conductivity BTU*in/(h * ft$^2$ * °F) | Thermal resistivity (h * ft$^2$ * °F)/BTU*in |
|---|---|---|---|---|---|
| A1 | 43.04 | 0.0430 | 1.0000 | 0.298 | 3.36 |
| A1 | 42.86 | 0.0428 | 1.0010 | 0.297 | 3.37 |
| A1 | 42.70 | 0.0428 | 0.9982 | 0.297 | 3.37 |
| A1 | 42.53 | 0.0429 | 0.9906 | 0.298 | 3.36 |
| A2 | 37.24 | 0.0406 | 0.9182 | 0.281 | 3.56 |
| A2 | 36.89 | 0.0405 | 0.9112 | 0.281 | 3.56 |
| A2 | 36.91 | 0.0406 | 0.9102 | 0.281 | 3.56 |
| A2 | 36.58 | 0.0408 | 0.8962 | 0.283 | 3.53 |
| A3 | 48.94 | 0.0462 | 1.060 | 0.320 | 3.12 |
| A3 | 48.74 | 0.0458 | 1.064 | 0.318 | 3.14 |
| A3 | 48.64 | 0.0458 | 1.063 | 0.317 | 3.15 |
| A3 | 48.48 | 0.0456 | 1.063 | 0.316 | 3.16 |

**Table 5.4:** Classes of thermal insulation and their efficiency factors in 40 × 90 mm wood-frame walls without any additional external insulation (SI units in parentheses).

| Class of insulation, the resistivity of insulation ((m$^2$K)/W) | Thermal – conduct. coefficient (W/(m · K)) | Nominal R-value of the wall ((m$^2$K)/W) | Mean R-value of the wall from 2D code ((m$^2$K)/W) | Reduction from nominal R-value (%) | Efficiency factor for insulation |
|---|---|---|---|---|---|
| 21.8 | 0.046 | 2.26 | 2.00 | 11.3 | 0.89 |
| 27.7 | 0.036 | 3.79 | 2.36 | 15.5 | 0.85 |
| 34.7 | 0.029 | 3.40 | 2.73 | 19.9 | 0.80 |
| 41.6 | 0.024 | 4.02 | 3.06 | 23.9 | 0.76 |

**Table 5.5:** Classes of thermal insulation and their efficiency factors in the 2 × 4 wood-frame walls with external thermal insulation: (a) R1 and (b) R1.6 (m$^2$K)/W.

| Class = resistivity ((m$^2$K)/W) | R-value of ext. insulation ((m$^2$K)/W) | Nominal R-value ((m$^2$K)/W) | Mean R-value from 2D code ((m$^2$K)/W) | Reduction from nominal percent | Efficiency of the insulation |
|---|---|---|---|---|---|
| 21.8 | 1.0 | 3.25 | 3.01 | 7.4 | 0.93 |
|  | 1.6 | 3.85 | 3.61 | 6.2 | 0.94 |
| 41.6 | 1.0 | 5.00 | 4.12 | 17.7 | 0.82 |
|  | 1.6 | 5.60 | 4.74 | 15.5 | 0.85 |

While external insulation is recommended for wood-frame walls, it is necessary for steel.

## 5.9 Thermal bridging in the steel stud walls

With steel being about 1,000 times more conductive than fiberglass, standard steel frame walls may be considered a thermal bridge filled with insulation. Figure 5.13, based on ORNL data, shows the efficiency of fiberglass in the steel frame wall to be below 50%. Codes and standards, therefore, require the use of exterior insulation. There is little difference between different fiberglass batts located in cavities of 100 or 150-mm nominal walls until R1 (m²K)/W exterior insulation is added. Yet, with an adequate level of exterior insulation, for example, R2 (m²K)/W, the efficiency of the fiberglass batt is better (Bambino, 1999).

Exterior gypsum sheathing is attached to the outside face of the steel frame, and an air barrier system is adhered to the gypsum sheathing with a minimum of 50 mm of extruded polystyrene placed on top of it as the exterior insulation; this wall assembly is likely to perform well.

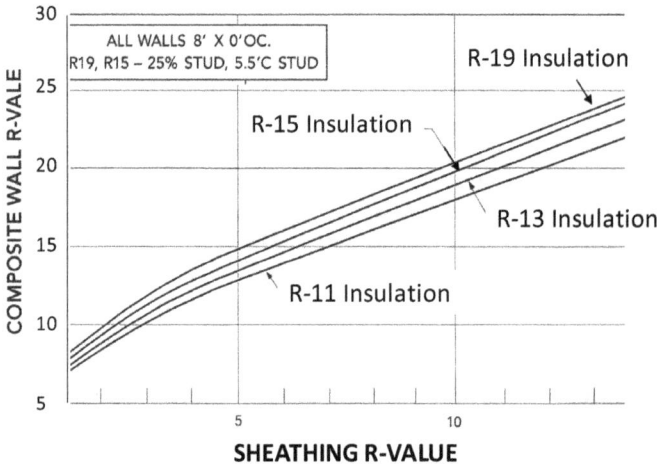

**Figure 5.13:** Approximate R-values in (ft² * °F * h)/BTU of steel frame walls with exterior insulating sheathing.

Steel studs using modified profiles to increase the length of the heat conduction path and perforated steel profiles were already known 65 years ago. Berthier (1960) reported on several tests; see Figure 5.14. Improvements to steel walls were intermittently noted in Canada (Sasaki, 1971; Brown, 1986) and in Finland (Nieminen and Salonvaara, 1999). The principle of parallel slots in the central section of the studs, extending beyond the end of each other, is well documented. In the Finnish (patented) steel profile, heat flux reduction through the thermal bridge was 40% to 50% compared to ordinary steel. In the best combination of thin profile and slots, the reduction

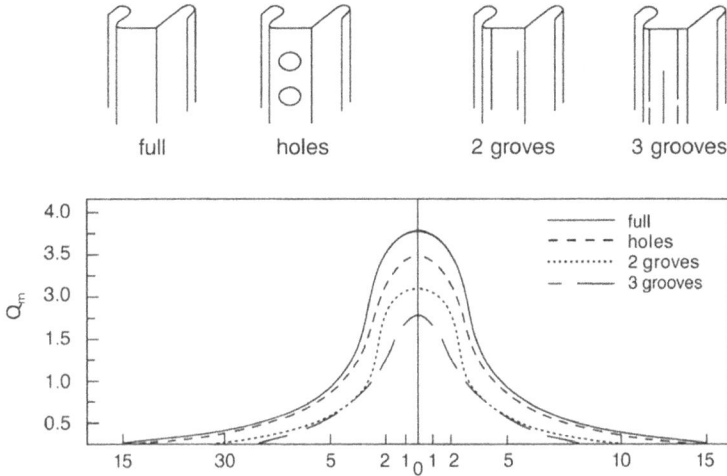

**Figure 5.14:** Comparison of thermal efficiency between four types of steel profiles (Berthier, 1960). Note: The picture is from the original publication, the English term for discontinuity in the metal is "a groove," while in North America it is "a slot."

can be as much as 85%.[2] Yet, with a few exceptions, only the complete, standard steel profiles are used in North America.

## 5.10 Thermal bridges and surface condensation

As previously highlighted, the depression of temperature on the inner surface of a thermal bridge may cause condensation. Often, temperature depression on the surface is reinforced by other factors, such as restricted air circulation. In one case, a baseboard heater was installed on one of the exterior walls. The mold growth pattern was modified but not eliminated. Why? The heater did not work continuously, and warm air pulled more vapor, so when it cooled again, it condensed on a wider area. To stop window condensation, one should use a much smaller but continually operated heating wire.

Condensation in the corner of exterior walls, one should have an exterior thermal insulation applied, and if not possible, the solutions on the interior side are:
1) Placing a heating wire (with a thermostat control) in the very line of the corner
2) Eliminating the corner by filling it to a curve to increase the heat transfer surface
3) As (2), but using a hygroscopic material that could act as a moisture buffer between night and day conditions
4) As (3), but using active capillary capability to remove condensed water
5) A combination of measures

---

2 Michael Salonvaara - private communication.

## 5.11 A closing puzzle

Figure 5.15 shows a picture taken from technical literature. The authors show a thermal bridge across R30 fiberglass batt (230 mm thick), highlighting the thermal shortcut and airflow through the thermal insulation. Yet, they did not mention that this construction is designed to fail because of condensation that may cause steel corrosion. Samuelson (1980, HRef) and Witt (1999, HRef) showed that this type of metal deck does not control air or vapor flows and that high RH will result in water vapor condensation on the top side of the metal deck.

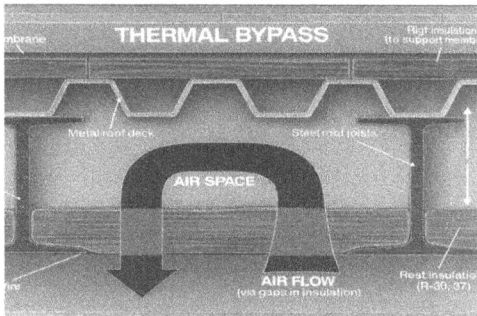

**Figure 5.15:** What is wrong with this design?

How should we design such a deck? Simple, there should be no thermal insulation under the deck. The air and vapor barrier should be applied to the substrate boards (exterior gypsum, OSB, plywood) placed on the top of the deck. On top of them, one builds an actual insulated roof (inverse, green, or standard low-slope roof). The temperature between the suspended (acoustic) ceiling and the roof deck should be near that of the conditioned space. This requirement stems from both reducing the risk of condensation and IAQ requirements. The space between structural beams should be used to carry HVAC ductwork, piping, and other services.

## 5.12 Closure of chapter

Thermal insulation modifies hydrothermal performance in different ways. External insulation mitigates the effects of thermal bridging and water vapor control. Nevertheless, if the air barrier material is impermeable to water vapor diffusion, one must consider drying on both sides of the wall assembly.

When dealing with thermal bridges, one must carefully evaluate the construction details for heat, air, and moisture. Special care must be taken when designing details such as steel column penetrations, supports for window-washing platforms, floor penetrations (Child, 1988, HRef), as well as the floor junctions in masonry construction or balconies connected to concrete decks because of the significant depression of the surface temperature and frequent occurrence of standing water in those junctions. Other thermal bridging considerations include cases where some reduction of convective air movement may accompany temperature depression (e.g., corner closet or the staircase leading to a roof terrace).

# References

Bomberg, M.; Lstiburek, J.W.. Field Applied Spray Polyurethane Foam Envelopes of Buildings. 1998, Technomic Publ. Co., pp. 1-339.

Bomberg, M.; Kisilewicz, T.; Mattock, C.. Methods of building physics. Cracow University Press, 2015, pp. 1–300.

Bombino, R.; Burnett, E.F.P.. Design Issues With Steel Stud-Framed Enclosure Wall Systems. Pen State University, PHRC, Res 58, 1999, pp. 44.

Berthier, J.;. Les points fables termiques ou ponts termiques. *Raport Centre Scientifique et Technique du batiment*. 1960, *42*, 36.

Ossowiecki, M. Curtain Walls (In Polish). Arkady, 1964.

# Chapter 6
# To test or to guess: paradoxes of testing within building science

Classification systems enhance communication within a specific discipline. Yet, the definitions and scope of building science had always been left to interpretation, as the only definition was made ad hoc with its objective. The scope of building science may have been identical to building physics, as the latter was defined by tradition. We add some terms from L. Bachman's architecture course at Houston University, 2013, who defined:
(1) Data are the measured characteristics of the material or system properties.
(2) Information is an effect of the process of transforming data into a form on which the decision process can be based.
(3) Knowledge is information that is consistent with other fields of organized data.
(4) Understanding is the ability to reproduce knowledge from first principles and apply it to unique situations.

The following hierarchy of the built environment separates different data fields:
–   Town (city).
–   District of the city.
–   Cluster of buildings (or building quarter) with some common technical features.
–   Building (considered as a system).
–   Building assembly (e.g., wall) or building subsystem (e.g., heating or ventilation).
–   Building component (window) or part of the subsystem (heat pump).
–   Construction composite (panel with several materials joined together).
–   Construction material (coating, membrane, or a commercial product).
–   Material matrix (a component of the structure).

Today, most professionals would agree that there is no schism between building science and physics, and all professionals strive toward predictable field performance. The analysis of building in cold climate (Bomberg et al., 2016) and this book defines the cluster of buildings as the basic level for technological development because we need to analyze solar and geothermal components as they interact with the building. A designer or an architect is asked to consider an active linkage with water tanks in the ground as the thermal storage medium and to integrate the building with the surrounding nature. A good example of this thinking is in Japan, where each piece of soil is used for placing greenery to break the concrete look of the city.

   To speak about the predictability of performance, we need to use modeling and testing. Modeling is necessary to account for the outdoor climate. It must be calibrated for a given structure and climatic conditions, and therefore, we have introduced the MAPE (monitoring and performance evaluation) block. Assuming monitoring to pro-

https://doi.org/10.1515/9783112217023-006

vide the required data, whatever model of hygrothermal or energy performance is used, after calibration, it can be used for steering heating/cooling, ventilation, and air conditioning the equipment.

Prof. Hugo Hens (K.U. Leuven), speaking at Westford, MA, seminar in 1998, defined building physics as:

– Applied science, a basic subject in building engineering education
– Containing sub-fields: heat and mass transfer, acoustics, light, and sometimes fire; performances of materials, building elements, and buildings, with a view of human comfort, health, service life, costs, and sustainability

This short description of the building physics domain shows how large the multidisciplinary field for performance evaluation is, in which no performance test methods exist. Over two decades, starting in the 1970s, the National Bureau of Standards led a significant international effort to create a system of performance-based test methods but failed to achieve any progress because of difficulties in their scope of validity. For example, if we use a bag of sand falling 1.5 m to hit drywall, would this also be a valid test if the drywall is mounted on vibration damper support? Could one repeat this test three times?

American Society for Testing and Materials (ASTM) uses comparable tests that provide either a physical or apparent characterization of an untreated material. The apparent or effective value from a test means that a result depends on the test method, for example, water absorption with the specimen keeping surface water during the weighing test. Thus, what can we do if we want to evaluate a material exposed to specific environmental or a long-term performance of a composite material?

We will also use an apparent test, but we must ensure that we have not simplified the conditions of the test by eliminating some of the critical factors. In this chapter, we will also provide you with some examples of apparent tests used by the authors.

## 6.1 Do not simplify before you understand, simplify test when you know all critical factors

Water in building materials was not a serious consideration as masonry dries after becoming wet by rain. Yet, water is a serious consideration when water vapor condenses in airtight wood-frame walls. Scientists have dealt with this problem since the 1930s. Yet, in 1958, when Glaser described a simple method to calculate condensation, water vapor condensation became a subject for public rationalization. Even though his model was oversimplified as water can evaporate and continue to move because of phase change, capillary and electro-osmotic forces, freezing pressures, etc., his model remained in the curriculum of many European schools for 50 years. Note it took 15 years for Scandinavian computer models to replace it and 35 years for professional German computer models to replace the oversimplification.

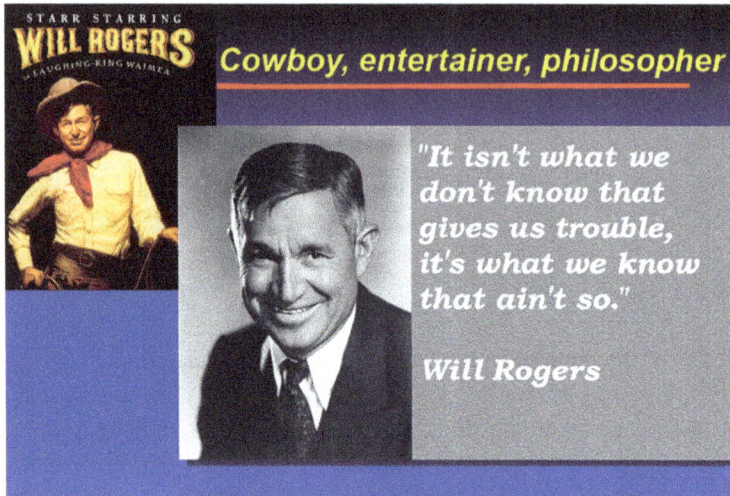

**Figure 6.1:** Will Rogers expressed the oversimplification like Einstein on being simple but not simplified (courtesy of Lew Harriman).

We bring the issue of Glaser model misuses to calculate the amount of condensation because teachers often use simplifications to explain complex phenomena. Typical simplifications in those courses are using temperature and relative humidity instead of thermodynamical potentials for energy or chemical processes.

Performance testing is particularly difficult as each of them must operate within a defined field of validity. Within ASTM, there is a special category of tests called "ruggedness test," where one provides a range of possible changes in boundary conditions for assumed factors affecting the outcome and a few dummy variables (total n variables), and when performing (n + 1) tests, one may see the difference in the impact of different factors.

This chapter reviews issues of testing, and we selected presentations in the form of paradoxes. According to the *Oxford American Dictionary* (1980), a paradox is a statement that seems to contradict itself or conflict with common sense, but which contains a truth.

## 6.2 Paradox of testing and knowledge

The paradox of testing and knowledge may be stated as follows:

> To design a test that provides a required judgment value e.g., assess field performance of an assembly or material, a broad knowledge is needed, but to gain this knowledge requires complex experimental work.

To perform such experimental work, one may use the following techniques:

(a) Block of test methods that demonstrate the effect of one factor in a few different situations
(b) Sensitivity analysis in an experiment or on a model of the phenomena to rate several effects of many factors
(c) Comparative tests of many materials to measure the response of each material tested under the same conditions

### 6.2.1 Block of testing (to demonstrate the effect of one factor in different exposures)

The example below demonstrates different thermal performances as environmental conditions change. There are two materials, either covered by an almost perfect water barrier with a few pinholes or with an open surface. The experiment's conclusion, shown in Figures 6.2 and 6.3, may surprise some people. First, the sealed specimen picked up a very small amount of water vapor compared to the materials with exposed surfaces, but it also dried much slower than those with exposed surfaces. Effectively, after 5–6 months, both open surface specimens were dry, while those covered by an almost perfect moisture barrier retained most of the initial absorbed water vapor.

The explanation is simple. The water vapor entering through a pinhole condenses, and the combined liquid and vapor transport mechanisms redistribute the small water quantity. Later, in the drying stage, only a small area of pinholes dries using diffusion, and the water remaining in the material must diffuse to the dry spot next to the pinhole. In contrast to this mechanism, a vast amount of water vapor enters through the open surface of the material, and the whole surface of the material takes part in the transport processes.

In the wetting part of the test (marked as Δt), the water vapor diffusion through the pinholes is governed by the size of the openings and is not affected by the type of material on the other side, so the starting point for drying of both specimens with five pinholes is almost the same. The drying rate is also similar because the water vapor permeabilities of both materials are similar. Conversely, the wetting part of the open surface specimen shows large differences between low and medium-density EPS. Type 2 (medium density) has a much higher holding power for water vapor than type 1. Type 2 collects much more water vapor and dries longer than type 1.

In conclusion, type 1 (low density) is better suited for wet applications than type 2 (medium density). Based on tradition, the trade people prefer to use the nominal

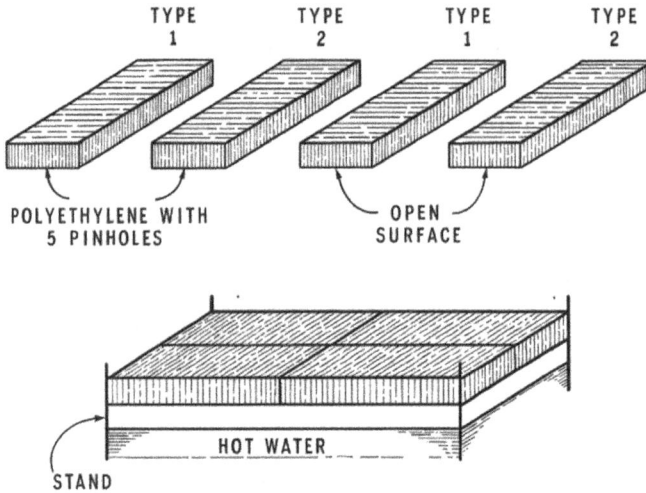

**Figure 6.2:** Block of testing for expanded polystyrene (EPS) with a density of 15 kg/m$^3$ (type 1) and 20 kg/m$^3$ (type 2), wrapped in a sealed polyethylene bag with five distributed pin holes and with open surfaces, placed above a container with 70 °C water.

15 kg/m$^3$ material in freezers and all applications that include exposure to water and where mechanical strength is not required.

This test demonstrated that exterior basement thermal insulation wrapped in polyethylene with a 30 cm (1 ft) overlap became soaked with water after a two-year exposure, while the same EPS without protection or closed-cell polyurethane was completely dry in the same exposure.

### 6.2.2  Sensitivity analysis (to rate effects of various factors in different exposures)

We demonstrate the effects of boundary conditions on water ingress into the thermal insulation materials and used the experimental set-up below to highlight the differences related to the exposure type. Three types of exposure were used:
1. Isothermal conditions
2. Constant temperatures on both sides of the specimen in the form of a slab
3. Variable temperature on one side and constant on the other side

The conditions of exposure are shown in Figure 6.4.

Comparing the results obtained under three exposures, one may observe each time a different order of the three tested materials: the highest water ingress to the low-density expanded polystyrene under isothermal conditions. Isothermal water ingress has the smallest driving force, and the large volume of air voids in this material facilitates the largest entry of water.

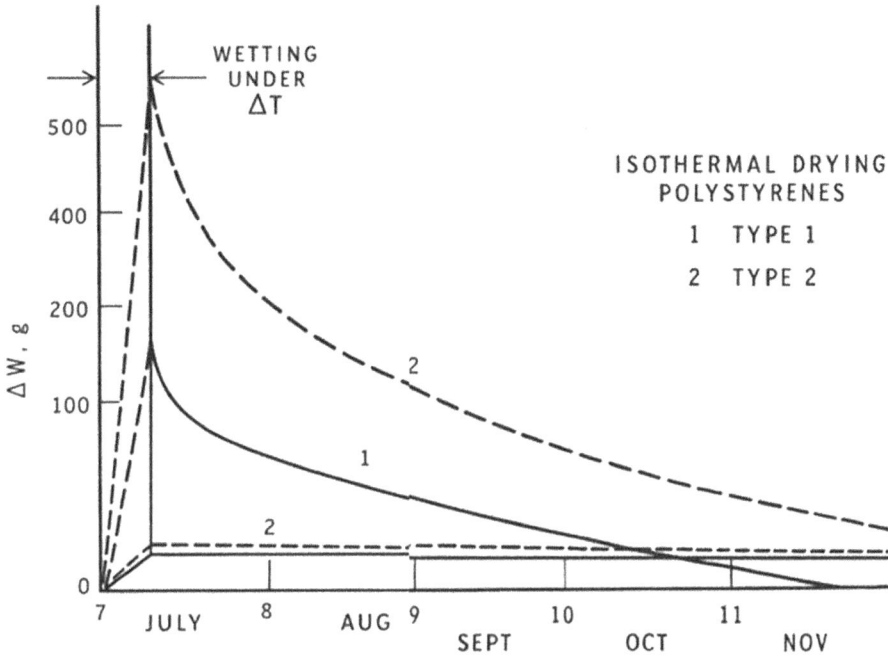

Figure 6.3: Initial increase in moisture content during 7 days of exposure to the source of water vapor, followed by the removal of the water tank and laboratory drying for 5 months (see text).

**1. ISOTHERMAL WATER INTAKE**

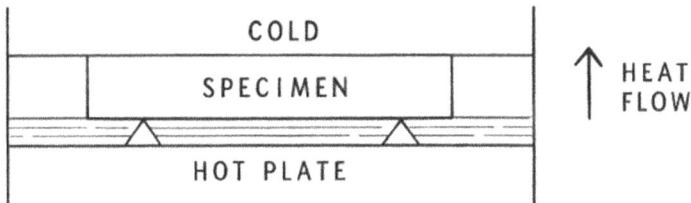

**2. WATER INTAKE UNDER ΔT**

**3. ΔT CONSTANT/VARIABLE FREEZE-THAW**

Figure 6.4: Three experimental set-ups for testing the effect of boundary conditions: (1) isothermal capillary uptake, (2) constant thermal gradient, (3) thermal gradient under freeze-thaw conditions.

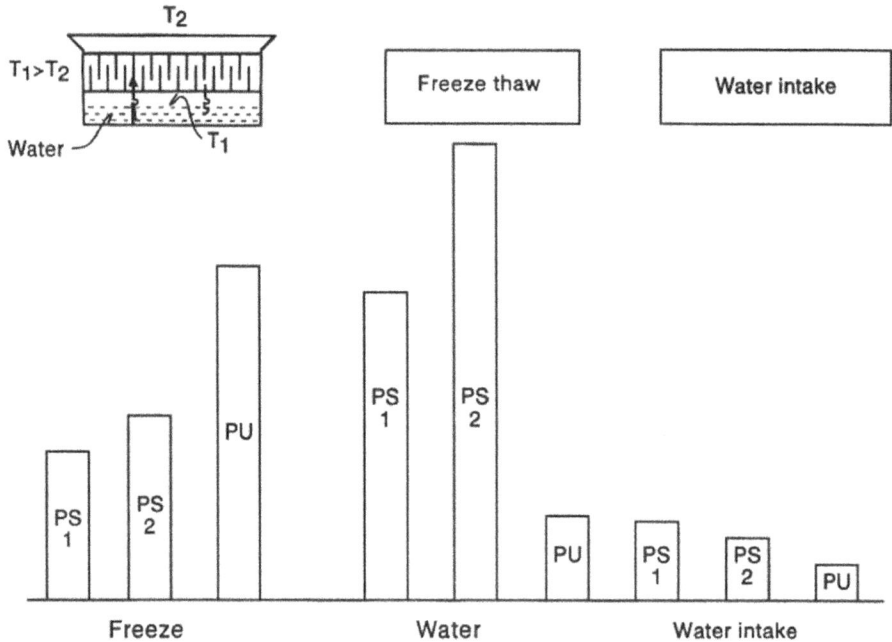

**Figure 6.5:** Results of the sensitivity test with boundary conditions are shown in Figure 6.4. Materials: (expanded) PS1 and PS2, and the medium-density spray polyurethane foam (PU).
Note: the bars are drawn to scale, but the scale does not have numbers because the volume of water attached to the specimen surface is unknown, and it varies depending on the nature of porosity.

Constant thermally driven water vapor transport loads more water than the variable (freeze-thaw) to the closed-cell polyurethane. There is no redistribution of water because the cells are closed. On the contrary, the medium-density expanded polystyrene acts like a cyclic sponge; water entry by thermo-diffusion is redistributed in open cells, and the next portion comes next day. This is why the material testing must relate to representative climatic conditions.

## 6.3 An appropriate test method to evaluate the performance of a material under service conditions

In this context, an appropriate test method means:
1. Precise (repeatable, reproducible, Figure 6.6)
2. Accurate (may be traceable to the national standard of the measured property)
3. Realistic, but also a simple test
4. Representative of the performance attribute being evaluated
5. Have a judgment value

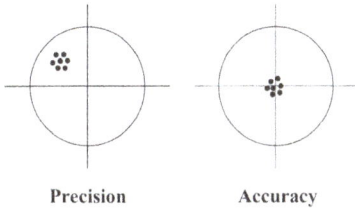

**Figure 6.6:** Visual representation of precision and accuracy.

Repeatability precision deals with the variability between test results obtained in one laboratory when the test method is applied to a given test specimen. Typically, it is expressed as the standard deviation of the test results multiplied by a factor related to the number of tests and the selected probability of the results.

Reproducibility precision deals with the variability between single test results obtained in different laboratories, each of which has applied the test method to the same test specimens.

Accuracy is reproducibility precision in relation to the internationally established definition of a physical property or to the transfer standard of the required physical property established by a recognized national laboratory, for example, the National

**Figure 6.7:** In this test protocol, the criterion is the time required to fill a specified commercial plastic bag. To make it a test procedure, one must specify repeatability precision within and between testing laboratories.

Physical Laboratory in the UK or the National Institute of Standards and Technology in the USA.

The phrase "to evaluate performance of the material"

(a) How well do the test conditions agree with the service conditions
(b) Selected pass/fail and number of the tests
  – Average of a specified number of repetitions
  – 95% probability (random selection of batches and size of each sample)
  – Using a ranking scale of performance

To illustrate the significance of sampling to produce a nominal value, compare thermal testing in Europe, Canada, and the USA. The test methods used for thermal conductivity determination are practically the same in all cases. In all cases, however, the sampling procedures are different.

ISO standard is the most precise and set limits concerning the sample size. The Canadian standard uses a mean value of three specimens, requiring it to be no less than the nominal value. All specimens can be at most 10% lower than the nominal value of the listed resistance. The US standard is based on a large number and requires that the average be equal to the nominal value of the listed thermal resistance. The published analysis showed that both Canada and ISO standards were within 1–2% of the real nominal (i.e., listed) thermal resistance. Still, the US products were 6–8% lower, while the specific numbers depended on each product's variance. While we do not discuss the implications, the above discussion highlights the significance of prescribing the sampling procedure when one talks about the precision of the test method.

The phrase "under typical service conditions"
Construction:
– type of the building, and
– type of building component.

Service conditions:
– indoor climate (inside the building),
– microclimate (within the building enclosure), and
– mezzoclimate (outside the building).

Figure 6.8 collects all these considerations into one graph. We do not discuss the process of establishing a nominal or representative value of the product; we are only interested in highlighting the number of considerations.

## 6.4 Paradox of material variability

One may formulate the following paradox of material variability:

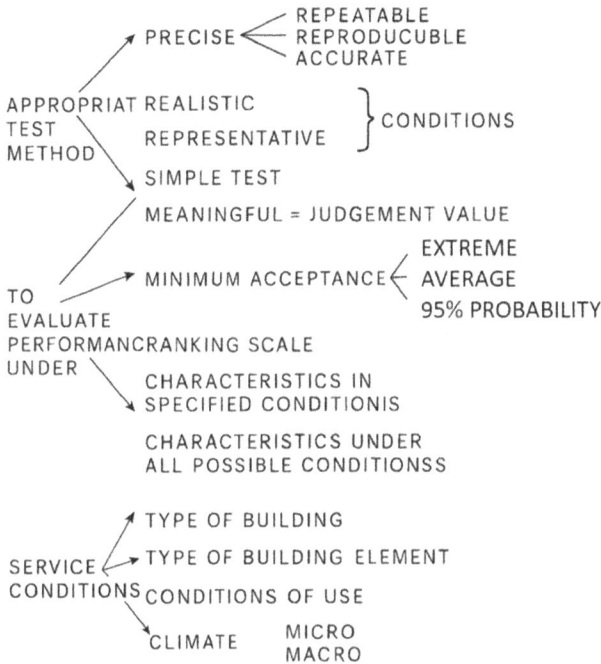

**Figure 6.8:** List of considerations in selecting an adequate field performance test.

to establish material variability in any test, one must ensure that critical factors affecting the test precision are included in the test specification.

We are saying that if some critical factors affecting the variation of the results are not included in the test protocol, the variation of the test results will be closer to random. The following experiment illustrates it.

Several unprotected types of exterior underground thermal insulation were exposed for 2 years to a silty clay backfill. The question was, "has the mechanical performance of the insulation been changed." Specimens used for thermal testing have dimensions of 30 × 30 cm, and those for compressive strength have dimensions of 4 × 4 cm. So, the laboratory doing the test that used density determination on 30 × 30 cm came to inconclusive results. The unbiased results required testing the density of 4 × 4 cm squares.

When this was done, linear graphs of compressive strength versus density before and after exposure crossed each other at the average density of the whole population, indicating that some structural changes took place, but overall compressive strength was retained.

## 6.5 Paradox of accelerated tests

With exception of surface phenomena, there are no methods known to accelerate weathering processes, *yet* we can change the impact of the selected factor to reduce the test period.

**DIFFUSION**

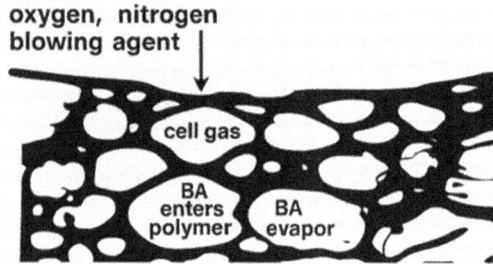

**Figure 6.9:** The initial stage in the closed-cell gas-filled foam, for example, polyurethane or matrix.

Figure 6.9 shows the initial state of thermal insulating foam, such as polyurethane or polyisocyanurate. Foam cells are filled with a low-thermal conductivity gas (blowing agent) and a co-blowing agent (if applied), initially there are no air components in the foam. Oxygen and nitrogen from the ambient air will now diffuse inward, changing the concentration of the blowing agent in the cell gas. During the foam growth, the cell gas pressure is slightly higher than atmospheric, and the pressure of the blowing agent drops to 0.6–0.7 atm after cooling. Assuming that the cured foam will be exposed to oscillating temperatures, the blowing agent condenses, evaporates, and in a smaller or larger extent, enters the solid polymer matrix.

All those processes together change the pressure and composition of the cell gas in the process called aging. While those processes are described in detail in a book by Bomberg and Lstiburek (1998), here we will only look at the means of eliminating material variability when defining the rate of aging.

One can use the theory of mechanical similitude, and for a diffusion process in a slab; to maintain the same conditions, one must keep the Fourier number constant. That means the ratio of the product (time and diffusivity coefficient) to the second power of the characteristic length must be constant, as follows:

$$Fo = (D_{eff}t) / L^2 \tag{6.1}$$

The acceleration of the diffusion process will be related to the second power of the slab thickness; see eq. (6.2), where the subscript denotes the reference material:

$$t / t_o = L^2 / L_o^2 \tag{6.2}$$

The above relationship indicates that properties for a thick material can be predicted from measurements on thin material with the same properties.

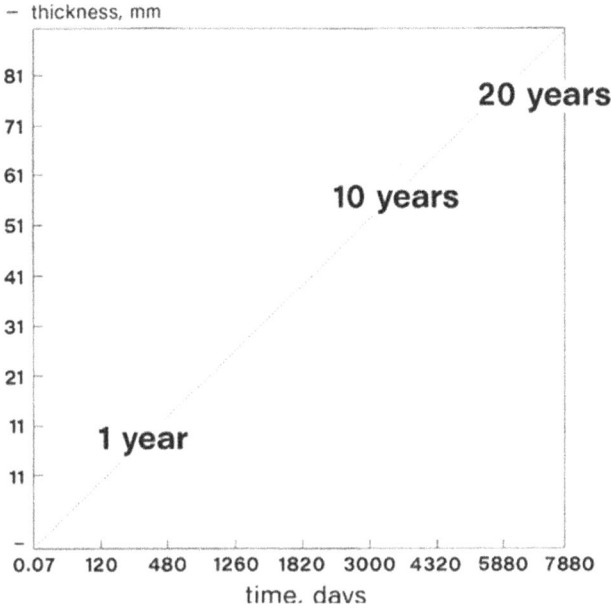

Figure 6.10: Relation between the diffusion period and the characteristic length (thickness of the slab) in diffusion with a constant Fourier number.

Thus, when a 50-mm-thick slab is represented by measurements performed on a 10-mm-thick layer, the exposure time is shortened from 3,000 to 120 days. We may use the definition of long-term thermal performance, that is, 15 years of exposure being represented as the value obtained after 5 years of aging, that is, after $5 \times 365 = 1,825$ days of aging.

For 50-mm-thick foam, if using a 10-mm-thick slice of the material, the same degree of aging will be obtained after 73 days of aging. Of course, we assume that the foam is uniform and homogeneous. If not, the measurements will be performed on slices cut from different locations, and the calculations will be performed with a computer model.

If one chooses a variable that may reduce material apparent variability and uses a dimensionless variable instead, one may easily see the effect of remaining variables. The characteristic value for normalization depends on the nature of the analyzed process; for example, for moisture content, it will be the degree of capillary saturation; for the aging of cellular plastic, it will be dimensionless thermal resistivity (reciprocal of the apparent thermal conductivity) of the foam to the value measured in the initial time after foaming. Figure 6.11 shows a normalized thermal conductivity (starting from 1.0) of a unique spray polyurethane formulation. This figure shows thermal resistivity versus aging time, with measurements performed on foams produced by two manufacturers using five different product batches, measured on specimens with thickness varying from 5 to 20 mm, recalculated to 10 mm thickness with eq. (6.2).

Realizing that the aging process includes the inward diffusion of oxygen and nitrogen from the surrounding air, phase changes, outward diffusion, and solubility of the blowing agent in the polymer matrix, this agreement is sufficient for practical evaluation of the aging. It also indicates that the specimen preparation process was well-specified and that, despite different initial thermal performances, all specimens showed similar aging characteristics.

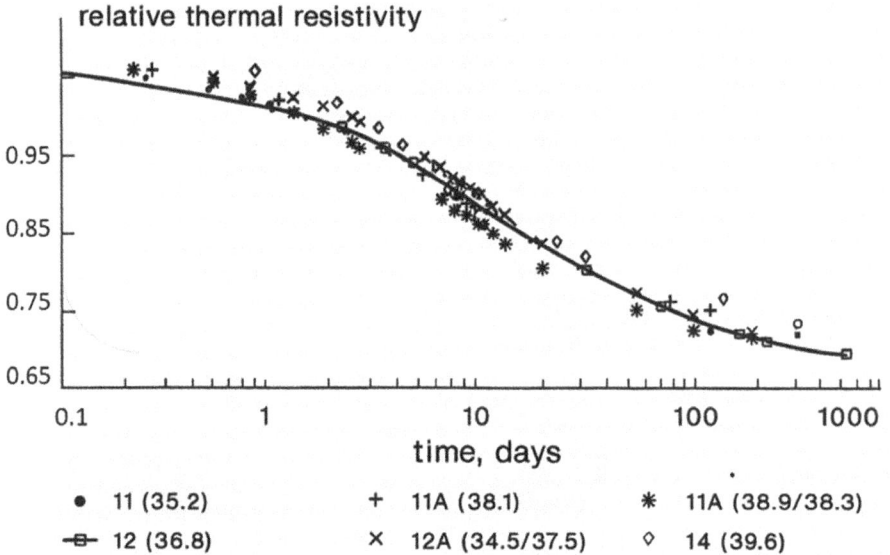

**Figure 6.11:** Dimensionless thermal resistivity ratio versus aging time.

## 6.6 Paradox of laboratory or field evaluation

The role of the initial and boundary conditions was clear when learning calculus. One can measure heat, air, and moisture flows. Knowing the boundary and initial conditions is necessary to be able to draw a conclusion. The best example is air tightness. The difference in the acceptance criterion of laboratory and field testing is 10-fold, namely (0.2 L/m$^2$s or 2 L/m$^2$s at 50 Pa pressure difference). We require 10 times tighter assembly for the laboratory test with known boundary conditions than in the field, where we do not know conditions at the boundary.

The paradox of laboratory and field testing may, therefore, be presented as:

Despite laboratory test being more precise, designer prefers testing construction assembly with a simplified test under field conditions because the latter case involve a better combination of the interacting factors.

Laboratory tests are to control if the manufacturer can produce an assembly, but because boundary conditions in the field are different, those tests have limited significance. One can observe that current trends are to increase the fraction of mock-up testing in the field rather than relying on laboratory evaluation.

## 6.7 Paradox of integrated monitoring, and modeling to replace testing

Modeling can be used to address the probability of specific weather or variability of material characteristics used as input to the hygrothermal models. Before computerized models, building physics was limited to understanding the observed failures and teaching how to avoid their recurrence in the future (Figure 6.12).

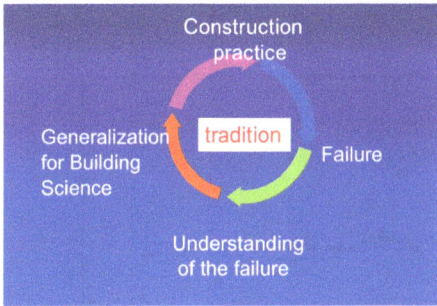

**Figure 6.12:** Traditional circle of progress in building physics was created by understanding the failure of real construction.

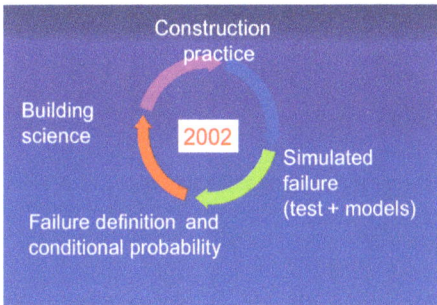

**Figure 6.13:** Building physics circle based on an integrated testing and modeling approach, where the simulated failure is derived from hygrothermal modeling.

Now, with modeling providing the definition of failure (Figure 6.13), building physics becomes a pro-active force. Nevertheless, each computer model includes several material properties (or characteristics), and their variability has a definite effect on the uncertainty of the model output. Calibrating a single hygrothermal model with another interacting model of air flows or CFD calculations has now given place to calibration on field-monitored data. Use of such data is necessary for the development of the next stage of field performance evaluation that we call MAPE monitoring and (field) performance evaluation.

We may formulate a paradox of monitoring, testing, and evaluation as follows:

"With monitoring building equipment performance in a specific location, used for model calibration and prediction for heating, cooling, and ventilation equipment performance, and tests leading to improved equipment performance, allow to modify models for predicting building system performance. Ultimately, we arrive at an integrated system to evaluate and improve energy and indoor environment performance, where models and tests create a continuum."

We can illustrate the application of integrated, computerized modeling to the evaluation of the long-term performance of a polyurethane composite foam system (Figure 6.14), where the assessment of risks introduced by the climate and material variability is clearly beyond the scope of experimental assessment.

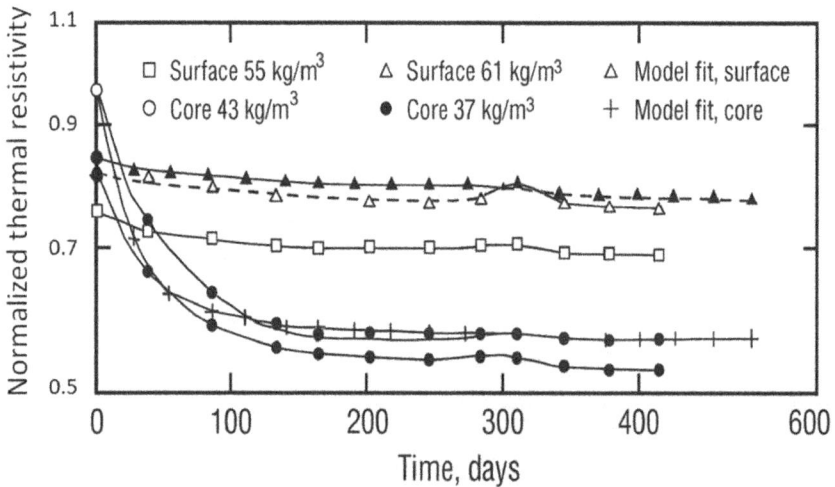

**Figure 6.14:** Comparing DIPAC model's prediction with the measured aging of the surface and core layers of non-homogeneous foam [30].

Modern polyurethane foams are non-homogeneous, with skins of higher density, as shown in Figure 6.14. Cutting thin layers from both surfaces and the core, and measuring the foam characteristic to enter them in DIPAC (distribution of integrated parameters, continuum) model to calculate the aging of a non-homogeneous foam. Figure 6.14 compares measured and calculated results on the surface layers alone. Additionally, Figure 6.14 compares different measurements on an actual roof (BUR) at increased temperature and kept in laboratory (agreement with DIPAC model calculations).

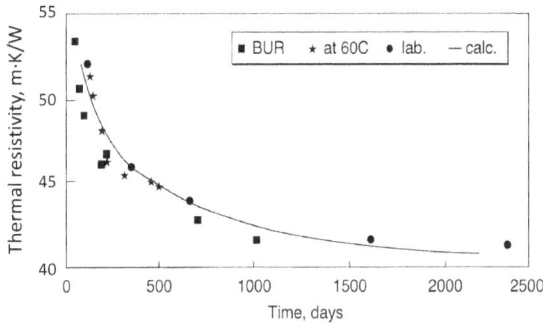

**Figure 6.15:** Comparing DIPAC model prediction (based on laboratory-determined material characteristics) and measurements on materials: (a) stored in the room, (b) stored in a 60 °C oven, (c) cut from a built-up roof (BUR).

## 6.8 Closure to chapter

This chapter talks about tools. We used the concept of paradoxes, as each of them brings to the reader's attention the existence of cross-effects and interacting phenomena. The outcome that appears like a contradiction is always an effect of a few different phenomena.

Observe that our test methods seldom relate to field performance because they were developed to compare different materials. For instance, clay bricks despite being used for the last 5,000 years, including at least 300 years of modern science. Do we know how to evaluate their durability? One says, talk about masonry because mortar repairs extend the life of bricks. The exterior plaster serves as weather protection, and the air cavity used in brick veneer is also a component of successful masonry. Thermal gradient that drives moisture depends on masonry type and geometry. So, the question should be asked about the cracking ability of a specific masonry system and not about bricks. Furthermore, masonry performance depends on weather and use conditions, and the question is re-phrased as the durability of masonry. A reply to this question is the spalling test under critical moisture content and a 1-D freeze-thaw cycle, in addition to the mortar-brick bonding under specified load and hygrothermal cycling. So, the answer about the durability of bricks involved two highly specialized methods of assembly testing.

We started the analysis with a logical approach; we found that the quest for simplification led us astray into material testing as much as in the human factors. We saw how, for the sake of economy and speed of construction, the materials were "improved," but the system was made worse. We saw how difficult it is to develop a test related to the field performance because the traditional methods are based on examining one factor at a time. We saw that for a system complexity, only a combination of testing and modeling can give us a path through the field of interacting variables.

Only the monitoring and modeling-for-performance evaluation system can be used for the evaluation for next generation of building technology.

# References

Bomberg, M.; Shirtliffe, C.J. Influence of Moisture and Moisture Gradients on Heat Transfer Through Porous Building Materials, Thermal transmission measurements of insulation. ASTM STP, Vol. 660, 1978, pp. 211.

Bomberg, M.; Lstiburek, J.W. Field applied Spray polyurethane foam envelopes of buildings. Technomic Publ. Co., 1998, pp. 1–339.

Bomberg, M.; Kisilewicz, T.; Mattock, C. Methods of building physics. Cracow University Press. see also the publication on the internet, 2016, pp. 1–300.

Bomberg, M.; Romanska-Zapala, A.; Yarbrough, D.W. History of American Building Science: steps leading to scientific revolution. *J. Energies.* 2020, *13*, 5, 1027.

Bomberg, M.; Romanska-Zapala, A.; Yarbrough, D. Towards a new paradigm for building science (building physics). *World.* 2021, *2*(2), 194–215. https://doi.org/10.3390/world2020013.

Hutcheon, N.B. The utility of building science. *J. Build. Phys.* 1971 1998, *22*, 4.

# Chapter 7
# Affordable, zero emission, environment retrofitting technology

So far, we have reviewed the issues of energy and environmental control and testing. In doing so, with a scientific point of view, we saw many technological elements that are suitable for the future. The following chapter is to integrate them into a universal retrofitting technology. As this project started exactly 20 years ago, when a program jointly sponsored by the NY State and the US Department of Energy ended without any further support, our microscopic, virtual network continued its work and published several papers. The first stage of the developed technology was called Environmental Quality Management (EQM, Bomberg et al., 2017; Romanska-Zapala et al., 2018, 2018b; Yarbrough et al., 2018). At this stage, we accepted that the starting point for development was the level of the US Passive House Institute. Therefore, as thermo-active insulation demonstration projects in Hungary and Japan fitted to our concepts. We decided to call our approach as passive and thermo-active cluster (PTAC) system technology. The word cluster is used here with two meanings: (1) as a cluster of buildings (any number between one with ground surroundings and two hundreds), and (2) as a cluster of thermal engineering ideas of passive or active type. One of these ideas is a problem-solving method that includes synectic and nature-like thinking.

## 7.1 Problem-solving method related to nature-like thinking

We use nature-like thinking that is analogous to eco-culture (Hays, 2017). We do not use solutions from nature, but we use the thinking paradigm typically used to describe nature. Often, it implies a direct connection between cause and effect without analysis of intermediate stages. Using this approach, we are not going to be concerned with the effect of the occupant's behavior on the efficiency of heating, but we will provide such a design that also gives some control of the indoor environment to the occupant. Of course, the design will be dual, optimized for cases when the occupant is not involved. Yet, the occupant is allowed to override the system if needed.

Similarly, we will drop the constant room temperature because adaptable room temperature is permitted in standards in America and Europe. Figure 7.1 shows the adaptable indoor comfort originated by the *Maximal Adaptability Model* by Hancock and Warm, discussing a broad range of comfort zones. What is remarkable is that over the whole optimal range of indoor temperature, the relative performance falls by only 4% (de Deer and Zhang, 2018).

Effectively, occupants do not feel discomfort if the temperature changes are slow, for example, 1 °C/h. We modify the indoor space at night's end and let the indoor tem-

https://doi.org/10.1515/9783112217023-007

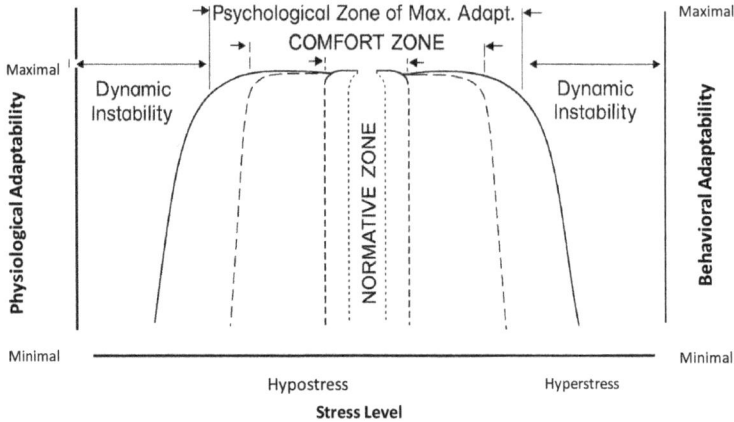

**Figure 7.1:** Relation between stress and adaptable comfort zone (De Deer and Zhang, 2018, from Bomberg et al., 2016).

perature come back to the limit of the comfort zone during the day. This type of thermal regime is beneficial for the morning period, acceptable during the day, and undefined for the evening period when occupants may set any preferred pattern. Moving the indoor temperature closer to the ambient conditions reduces heat transfer during the day and increases it at night. Shifting a fraction of energy from day to night will provide benefits to society.

As in retrofitting projects, one seldom has access to geothermal energy – at the best, one uses water tanks located either in the ground. For a low-energy house, passive house technology has reached the stage of maturity. For 70 kWh/(m$^2$·a), the heat transfer through the exterior enclosure is about 35 kWh/(m$^2$·a). Thus, if making a 20% reduction in the overall U-value, the total energy use is reduced only by 10%, implying that a new energy-generating method is needed.

## 7.2 Looking for a new energy-generating method

Mexican termites (Figure 7.2) with mounds on trees cannot use underground cooling like those living in Africa. Still, their mound is built with an axis indicating a specific time in the afternoon. A change in sun operation will make half of the mound exposed to early sun radiation, while the other half will be warmer in the late afternoon.

Figure 7.3 shows a wet, porous building material sealed in an impermeable bag and exposed to a constant temperature gradient. The heat flux (i.e., rate of heat flow) entering and leaving the specimen was measured. One can observe a large change in the rate of heat transfer at zero time, when the direction of heat transfer was changed to the opposite direction (Bomberg and Shirtliffe, 1978).

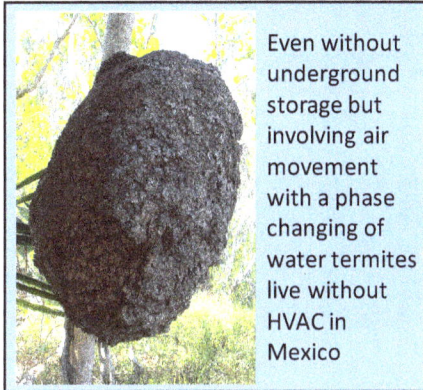

Even without underground storage but involving air movement with a phase changing of water termites live without HVAC in Mexico

**Figure 7.2:** Termite mounds in Mexico use natural cooling.

Figure 7.3 (reproduced before as Figure 6.2) explains how termites, by using high relative humidity and organic fibers with a significant fraction of nanopores for building their mound, can use a thermal gradient between material exposed to the sun and that in the shade for cooling. The queen resides in a central part that is cooled all the time.

A phenomenon called "thermal siphon" is created when a thermal gradient drives a vapor of any fluid, for example, water, toward the cooler temperature until it condenses. Then, the capillary forces transport the liquid phase of water back. The nanopores in the material will catch water molecules from the air during the nighttime, and mounds are prepared for the next day.

Figures 7.2 and 7.3 are provided here to visualize the heat siphon because, in all mechanical systems, the water pump moves the liquid back to the evaporation place. Like in nature, where different porosity of the material results in various rates of change, in the industrial situation, water-sourced heat pump (WSHP), that is, a mechanical device based on the heat siphon principle, uses different combinations of temperature and pressure applied to refrigerant gases to achieve cooling or heating. The WSHP is selected as an energy-producing source in our technology proposal; see below.

## 7.3 Passive and thermo-active cluster: a universal and affordable retrofitting technology

The PTAC involves a design, construction, and post-construction optimization of the many factors shaping energy efficiency and the indoor environment in new or renovated buildings (Bomberg et al., 2021, 2022). It involves using hydronic heating/cooling inside exterior or interior walls operating with temperature ramp 5–6 °C during the day (De Deer and Zhang, 2018). The energy is delivered during the night and stored in a water tank. One may ask a question why thermostat setbacks did not

Heat flux through sealed, wet specimen, water goes to the
cold side (at time = 0, specimen is flipped)

| | SIDE | MEAN.TEMP |
|---|---|---|
| • | HOT | 33.7°C |
| • | COLD | 14.25°C |

SPECIMEN
REVERSED,
HOT SIDE TO
COLD PLATE

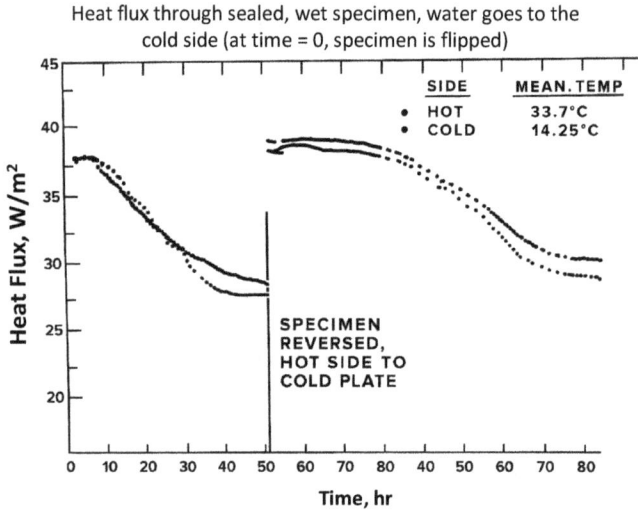

**Figure 7.3:** Demonstration of thermal siphon phenomenon on wet, porous material sealed in an impermeable box and placed in a heat flow meter apparatus that measures heat flux as a function of time (Bomberg and Shirtliffe, 1978).

work, and adaptable indoor climates work. The rate of temperature changes is 1 °C/h and fast on return, and it involves a different combination of climatic factors than a thermostat set back.

Work on PTAC started with a project co-sponsored by the NY State and the US DOE (Bomberg et al., 2009) and continued in China (Bomberg, 2010; Hu et al., 2013) as well as in the USA (Thorsell and Bomberg, 2011; Bomberg et al., 2015, 2016). New technology was the outcome of the works by Bomberg et al. (2016, 2017), Romanska-Zapala et al. (2018a, 2018b, 2019), and Yarbrough et al. (2018, 2019), and verified on masonry systems in 2024 (yet unpublished). Multistage process, described below, was demonstrated in Montreal, Quebec, Canada, in years 2008 to 2018.

### 7.3.1 Alleviating economic conflict with the two-stage (multistage) construction process

Investors must follow the requirements of codes and standards. Society, however, needs a higher investment level, net-zero energy, or at least near-zero energy. A two-stage construction process alleviates this conflict.

A multi-stage construction process, applied to a building cluster called "Atelier Rosemount" in Montreal, included a mix of different dwelling types, from inexpensive social to top-priced ecological dwellings. All improvements, except for photovoltaic panels, resulted in 64% and with photovoltaic we reached the total energy at 92% reduction. This project is interesting for two reasons: (a) it erased the difference between new and existing building; and (b) it shows that to achieve a 90% level of energy saving, the last 16% of energy must be brought by a new method of energy supply. These levels of energy reduction were confirmed in several Building America projects all over the USA.

### 7.3.2 Energy generation and storage units

In developing PTAC technology, one must observe two points. First, each dwelling in a housing block has a different demand, depending on its relative location (at the end or in the middle of the block) or difference in height to the neutral plane of air pressure. Typically, the lower half of the tall building has air pressure lower, and the upper half has the pressure higher than atmospheric. While easy to disregard in winter, this effect may be significant in the summer overheating. Thus, each dwelling must be treated as an independent energy unit.

While the use of water-sourced heat pumps (WSHPs) provides a higher coefficient of performance (COP), it also introduces one new problem, namely the need for an additional energy supply when the water temperature in the cold tank falls below a control point close to the freezing point, for example, 10 °C. To meet these challenges, PTAC technology introduced two concepts for thermal storage. One is a short-term thermal storage. It has 14–18-h thermal capacity to equalize daily loads and accommodate the requirements of the electrical grid. If the thermal storage of the building fabric is insufficient, one adds water in wall tubing and a water tank in the dwelling. The second is long-term storage, with weekly equalization of thermal loads and 168 h of weather variability. Typically, there will be a water tank located in the ground or elsewhere, and the extreme 5 days of cold weather are used for calculating this tank's capacity. Observe that in PTAC technology, we do not talk about equalization for periods longer than one week.

Heat pumps became indispensable because of their energy multiplication effect. Yet, the traditional geothermal heat pump is often not economic, and PTAC technology uses a cold-water tank in addition to a domestic hot water tank. A heating coil inserted in the domestic water tank and provided with a water pump delivers hot water to the floor and wall heat exchangers, and the water tank itself becomes the heat capacity provider. Nevertheless, as heat is continually extracted from the cold-water tank, and one must provide a source of energy to keep its temperature above the prescribed limit. This can be achieved in one of the three ways:

(a) Electrical heating of water (typically used)
(b) Use of a small geothermal or split air-water heat pump (used in large heating systems)
(c) Use of hot water transfer, typically when water recycling is in place (see below)

As the electric heating has an apparent COP of 1, as opposed to 4 for a WSHP, one uses a transfer procedure using a gray water tank (GWT). The transfer has two steps. In the first, one removes a specified volume of water from a cold water tank (CWT) to an auxiliary water tank (AWT) that should be located on a wall near the ceiling to allow gravity flow for the transfer to the gray water tank. In the second step, one fills the AWT with hot water from the domestic hot water (DHW) tank with the water pump 3 and delivers by gravity to the CWT. We do not specify the volumes of the DHW, CWT, GWT, or AWT because, generally GWT and CWT have the same capacity, but in retrofitting projects, their capacity depends on available space and permitted loads. And GWT depends on rain-water collection. AWT is typically one-fourth of the DHW tank. It may look like a paradox but adding a gray water tank lowers the volume of water delivered and removed and combined with optimization of COP making the system operation less expensive.

In a new building construction, one designs the DHW size for allowing 14–18 h use of the system's energy storage, that is, a short-term thermal storage period when the water system, together with the thermal mass contribution, eliminates WSHP operation during the day and daily peak loads. Furthermore, while the WSHP power is selected for 10 h of operation daily, the capacity of DHW must be sufficient to cover the difference between the average and 168 h of extreme cold weather (long-term thermal storage).

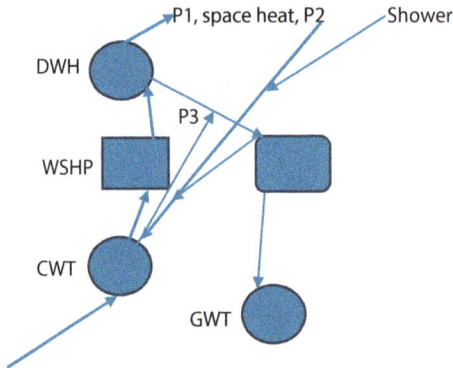

**Figure 7.4:** Schematic of connections (marked with thick lines) between the heating unit (CWT, WSHP, and DTH) and the system for hot water supply to the CWT (marked with thin lines).

Figure 7.4 shows the energy generation and storage unit (in short, energy unit) that is a self-contained, independent unit. It includes a CWT being a low terminal of a WSHP, and a DHW tank. The DHW tank has a water pump 1 that supplies hot water to the space heat exchangers, and if the water is not sent to the next building (see layer), pump 2 returns the water to the CWT. Both the CWT and DHW have connections (through pump 3) to an AWT. Water pump 3 and a three-way valve is located on the way from the CWT to the AWT and further to GWT. The AWT has an automatic contact that stops filling. Finally, there are two three-way valves with switches on the return from the space heating to the CWT (pump 2). One is from showers and the second from the AWT when hot water is delivered to the CWT (by gravity from the GWT). Thus, Figure 7.4 shows that adding a GWT and automatic filling the cold water tank includes a lot of automatics. For small domestic operation not using the GWT, placement of the AWT directly on line of pump 2 to heat water during the night and use it for heating the CWT in the day will be the preferred option.

### 7.3.3 Wall climate controlling unit

We decided to use the nominal temperature of the heating system for domestic use, about 50 °C depending on climate and efficiency of solar panels.

**Table 7.1:** Effect of the location of the radiant panel on energy demand in dynamic operation mode.

|  | Heating demand GJ | Cooling demand GJ |
| --- | --- | --- |
| Wall | 58 | 24 |
| Floor | 98 | 31 |

These values were calculated using Energy Plus computer software with typical film coefficients for horizontal and vertical orientations. Table 7.1 shows for heating one needs 60% and for cooling 80% of the floor area. Hu (2013) found that to achieve 90% efficiency in one-directional heating, one should place the PEX tubing on thermal insulation with a thermal resistance of no less than 1 (m$^2$ K)/W. Furthermore, using adaptable indoor temperature adds initiates an effect of mass on daily temperature variation (Fadeyev et al., 2017). So, we shall use a benchmark of 70% of the floor area for wall system heating. Generally, this excludes the kitchen and bathroom, but most of the rooms should have a wall heating system.

There are two methods of heat exchanger construction. In an in-situ application, the PEX tubing is continuous. In a panel application, the PEX tubing connects with snap-on joints. Panels can be made with different materials; for instance, in China, panels were produced with MgO cement and milled (fiberized) bush of rice and wood waste to provide both moisture buffering and elasticity.

## 7.4 An invention of a district climatic system to replace the district heating

This PTAC technology (Figure 7.5) introduces an important innovation in the form *of the climatic district network (CDN)* that may also be applied to historic buildings by pairing them with an adjacent standard building (Bomberg et al., 2011). In this manner, the pair of buildings can be included in a local district heating/cooling system. CDNs that include heating, cooling, and ventilation systems will easily eliminate the difference between a single building and district of the city. As PTAC technology includes thermal storage, and the water tanks may be located underground, the district climatic system may either be a part of the building itself or the energy-distributing system company.

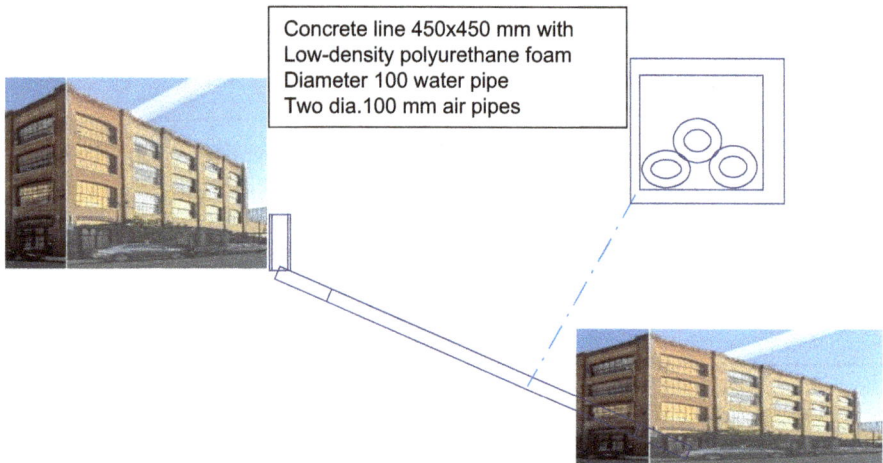

Concrete line 450x450 mm with
Low-density polyurethane foam
Diameter 100 water pipe
Two dia.100 mm air pipes

**Figure 7.5:** Schematic of the district climatic network. The underground, concrete line connects two buildings about 1 m below the ground surface.

The summary of the research on air-earth heat exchangers (Romanska-Zapala 2018b, 2019) is the basis of our proposal for 1 m depth in Central Europe. If a low-density (about 10 kg/m$^3$), polyurethane foam fills this underground line, the foam will be a dry insulation in winter and will become wet in summer, heat conductor in summer (gradient inwards) and as such it will dissipate heat better than a dry insulator.

In summary, sending return water, together with preconditioned air, to the next building may reduce operational costs by increasing COP over that of the heat pump alone. Furthermore, as the construction of settlements is less expensive than that of separate buildings, adding climatic district system increases investment efficiency and would diminish the difference between separate buildings and city districts (Yarbrough et al, 2019).

## 7.5 The benefits of the holistic approach to construction

The benefits of a holistic approach are as follows:
–    There are two methods of system construction:

(1)   in situ continuous construction of hydronic loops, or

(2)   a panelized system with heating and cooling, with or without ventilation.

–   Monitoring and modeling for evaluation (MAPE) is needed because monitoring field data (Klõšeiko, P., Kalamees, T., 2016, 2018) can be used to calibrate numerical models (Heibati et al., 2019, 2021) allowing their use for real-time calculation, as well as to optimize mechanical devices and improve the synergy of subsystems. Particularly successful are simple and precise artificial neural network models (Dudek et al., 2019; Dudzik et al., 2020a, 2020b; Bomberg et al., 2021).

–   Most of the mid-rises and high-rises use the corridor air pressure correction for the stack effect, which is different on each floor and varies with the seasons. Those differences have a significant effect on indoor air quality. Adding air pressure differences makes a small difference in monitoring cost but a large difference in the analysis capability, particularly when a variable rate of air exchange is included in the building design.

–   In a PTAC system, energy efficiency depends on keeping such boundary conditions as necessary to maintain the COP 4.0. The complexity and interactions of subsystems, particularly the MAPE system for multi-floor residential buildings, require performing a run-in period to check if the estimated energy efficiency is achieved.

–   PTAC also introduces a new type of climatic district network (Bomberg et al., 2021). Return water from building one is used for the lower terminal of the heat pump used in building two. CDN also includes preconditioning of air and eliminates the difference between designing one or several buildings.

–   Finally, the MAPE analysis should include both ventilation and unpredicted air exchange with the environment.

## 7.6  Summary of PTAC technology

The PTAC is proposed for universal public domain technology that can modify climate change. A water-sourced heat pump, together with two or more buffering tanks is the main heating or cooling subsystem. Figure 7.7 displays key elements of the water management system suitable for a hot climate. Water leaving wall exchangers is directed to the roof's surface before reaching a hot or gray water tank.

For summer cooling, one uses the connection from water pump number 2. When floor cooling is not sufficient, additional cooling will be provided by the ventilation system. Finally, water pump 2 will also supply cold water to the solar panels if they are connected to the house water management system. Normally, we will be using independent solar heating panels that are capable of self-orienting themselves toward the sun. Still, in the case of built-in solar panels, the hot water returning from the solar panels may be incorporated in DHW.

The ventilation air, taken from the air intake located next to the previous building, is also delivered by a climatic district network. The preconditioned air temperature may be too cold or too hot for the actual need a second air intake is recommended at building two. The operating system will select tone source that is closer to the need and let it go through the filtration and humidity control. Then, the air goes through the air handling unit (AHU) to a corridor ceiling plenum that is kept under a few Pascals higher pressure than the room. Air is carried in a large duct (minimum 150 mm in diameter or a side of the square) located in the plenum, and short flexible connections lead it to each ventilated room.

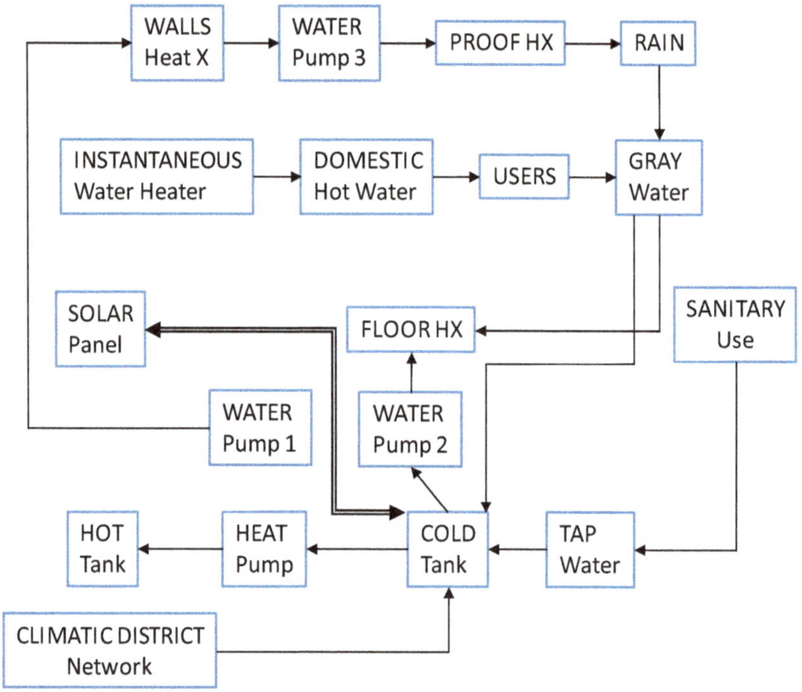

**Figure 7.6:** The water management system connects the exterior (climatic district network) and the interior (water collection to the gray tank).

The ventilation air, taken from the air intake located next to the previous building, is also delivered by a climatic district network. The preconditioned air temperature may be too cold or too hot for the actual need a second air intake is recommended at building two. The operating system will select one source that is closer to the need and let it go through the filtration and humidity control. Then, the air goes through the AHU to a corridor ceiling plenum that is kept under a few pascals of higher pressure than the room. Air is carried in a large duct (minimum 150 mm in diameter or a side of the square) located in the plenum, and short flexible connections lead it to each ventilated room.

This Canadian project demonstrated seamless thinking in construction in which the boundary between new construction and the retrofit of old buildings does not

**Table 7.2:** Cumulative reduction in the stepwise retrofit of Atelier Rosemount.

| The retrofit included the following steps: | Cumulative reduction |
|---|---|
| – High-performance enclosure; common water loop; solar wall | – 36% |
| – Gray water power, the cumulative energy reduction grows to | – 42% |
| – Heat pump heating All passive measures give a | – 60% |
| – Domestic hot water from evacuated solar panels | – 74% |
| – Photovoltaic panels reduce the total energy to a total of | – 92% |

exist. The Atelier Rosemount project broke another barrier, namely, the barrier of affordability. In the construction quarter, surrounded by four crossing streets, luxury apartments with windows on both sides of the building to enable cross-ventilation are neighbors of the social dwellings, with a cost being a fraction of the previous one. We will show the principle of financing on an arbitrary example. The building cooperative used the financing process of self-financing with increased mortgage payments from energy saving. Assuming that with stage one built, the estimate to reach zero energy is a US$50,000 loan per dwelling. People usually take a long-term mortgage on the full amount, and asking US$50,000 on a 6% and 25-year simple mortgage costs you US$46,645 with a total payback of US$96,645, almost double the loan. *Thus, a different financing approach must be used.*

Assume that you have additional financing ability of US$4,000.00 for years and 2,500 in the next 4 years, that is, total $18,000.00 and the current heating/cooling bill is US$330 a month or US$3,960 a year. You want to upgrade to zero energy over 3 years, in which the first retrofit stage costs you US$30,000, giving you after 3 years saving of US$150.00 a month. So, you need a loan for the first stage US$30,000, and you can pay US$350 from the extra money plus US$330 of the old mortgages, that is, 680 dollars a month. Looking at loan tables, you find 30,000 dollars paid in 4 years at 6% require a monthly cost of 705 dollars. You must pay US$21,868 from the principal and 3,768 dollars in interest charges in 3 years. At that stage, you will need to remortgage the property for another US$30,000 to cover the 8,100 dollars from the first stage and US$20,000 from the second stage. The cost of the old mortgage was about US$24,000, and the interest on the two loans was about 7,500 dollars. The contribution from energy savings was about US$6,000 from the first period and US$12,000 from the second so with US$44,000 paid for principal and US$8,000 for interest (i.e., US$52,000 spent), the cash of US$34,000 and the additional investment of US$18,000 gave you a total value of US$50,000. Your mortgage cost was about US$2,000 for 6 years of loans at 6% interest.

If you do not have extra money, you may use the same approach with an increased number of refinancing stages while keeping your initial monthly payments; yet, the process takes longer. In the case of Montreal building cluster, it took five stages and ten years of energy savings to achieve a 92% reduction in energy.

Currently, many financial institutions provide a loan based on existing capital resources. All a homeowner needs is a detailed cost of stage two for investment in reducing energy expenses. As lenders calculate a borrower's expenses as Principal, Interest, Taxes, and Insurance (PITI), adding energy bills to their calculation (PITI + E) opens an attractive investment opportunity. Observe that this investment does not increase the cost of living; the local economy gets a boost, while $CO_2$ emissions are dramatically reduced. This is a win-win-win situation.

A substantial part of this review was given to the gap between the advanced level of building science in the 1980s and construction practice. The first passive house conceived in the US and build in Canada in 1977 exemplified the problems created by traditional, fragmentary thinking. One tends to think that it was a past and we have learned. Thus, we report that in 2024, in the cold regions of Europe where the energy recovery ventilator was combined with typical AHU and the manufacturers eliminated preheating of outdoor air. The effect has been freezing of the water from outgoing air in severe cold climates. To eliminate water freezing in the AHU, we added to PTAC technology the automatic preheating of air by adding the air pipe to those of water in the climatic district network.

Today, in some regions, water management is becoming a growing issue, and therefore we expanded PTAC technology to include water recycling (gray water) to address energy and water savings at the same time. We highlight those details to explain how introducing a unified basic technology will bring many solutions to fall in place and bring cost reduction because of the increased volume of work.

# References

Bomberg, M.; Shirtliffe, C.J. Influence of moisture and moisture gradients on heat transfer through porous building materials. In Thermal Transmission Meas. of Insulation, ASTM STP 660, 1978, ed. Tye, pp. 211–233.

Bomberg, M.; Onysko, D., (Eds.) Energy Efficiency and Durability of Buildings at the Crossroads. 2008. 2008. http://thebestconference.org/BEST1 (accessed on 25 February 2020).

Bomberg, M.; Brennan, T.; Henderson, H.; Stack, K.; Wallburger, A.; Zhang, J. High Environmental Performance (HEP), residential housing and building technology, for NY state. *A Final Report to NY State Eergy Research and Development Agency and Nat.l Center of Energy Mgmt.*, manuscript, 2009, USA, 2009.

Bomberg, M. A concept of capillary active, dynamic insulation integrated with heating, cooling and ventilation, air conditioning system. *Front. Archit. Civ. Eng. China.* 2010, *4*, 431–437.

Bomberg, M.; Gibson, M.; Zhang, J. A concept of integrated environmental approach for building upgrades and new construction: Pt 1 – setting the stage. *J.Build. Phy.* 2015 2015, *38*(4), 360–385.

Bomberg, M.; Wojcik, R.; Piotrowski, Z. A concept of integrated environmental approach, part 2: Integrated approach to rehabilitation. *J. Build. Phys.* 2016a, 39, 482–502.

Bomberg, M.; Kisilewicz, T.; Nowak, K. Is there an optimum range of airtightness for a bldg? *J. Bldg. Phys.* 2016, *39*, 395–420.

Bomberg, M.; Yarbrough, D.; Furtak, M. Buildings with environmental quality management (EQM), part 1: Designing multi-functional construction materials. *J. Build. Phys.* 2017, *41*, 193–208.

Bomberg, M.; Romanska-Zapala, A.; Yarbrough, D.W. History of American Building Science: steps leading to scientific revolution. *J. Energies*. 1027 2020, *13*, 5.

Bomberg, M.; Romanska-Zapala, A.; Yarbrough, D. Towards a new paradigm for building science (building physics). *World*. 2021, *2*(2), 194–215. https://doi.org/10.3390/world2020013.

Building America. https://www.energy.gov/eere/buildings/articles/building-america-program

De Deer, R.; Zhang, F. Dynamic environnent, adaptive comfort, and cognitive performance. In Proceedings of the 7th International Building Physics Conference, IBPC2018, Syracuse, NY, USA, 23–26 September 2018; pp. 1–6

Dudek, P.; Górny, M.; Czarniecka, L.; Romanska-Zapała, A. IT system for supporting the decision-making process in integrated control systems for energy efficient buildings. In Proceedings of the 5th Anniversary of World Multidisciplinary Civil Engineering-Architecture-Urban Planning Symposium – WMCAUS 2020, 2020, Prague, Czechia.

Dudzik, M.; Romanska-Zapala, A.; Bomberg, M. A neural network for monitoring and characterization of bldgs with Environmental Quality Management, Part 1: Verification under steady state conditions. *Energies*. 2020, *13*, 3469.

Dudzik, M. Toward characterization of indoor environment in smart buildings; Part 1: Using the Predicted Mean Vote criterion. *Sustainability*. 220, (12), 6749.

Fadiejev, J.; Simonson, R.; Kurnitski, J.; Bomberg, M. Thermal mass, and energy recovery utilization for peak load reduction. *Energy Proceedia*. 2017, *132*, 38.

Heibati, R.S.; Maref, W.; Saber, H.H. Assessing the energy and indoor air quality performance for a three-story bldg using an integrated model, Pt 1: The need for integration. *Energies*. 2019, 2019, *12*(24), 4775.

Heibati, R.S.; Maref, W.; Saber, H.H. Assessing the energy, indoor air quality, and moisture performance for a three-story building using an integrated model, part three: development of integrated model and applications. *Energies*. 2021, *14*(18), 2021, 5648. https://doi.org/10.3390/en141856481.

Hu, X.; Shi, X.; Bomberg, M. Radiant heating/cooling on interior walls for thermal upgrade of existing residential buildings in China. In Proc. of the In-Build Conference, Cracow TU, Cracow, Poland, 17 July 2013, pp. 314.

IDP, 1991. An Integrated Approach to Design of Protocol Specifications Using Protocol Validation and Synthesis. *IEEE Trans. Comput*, Apr. 1991, 459–467, 40. doi: 10.1109/12.88465.

Kisilewiicz, T.; Fedorczak-Cisak, M.; Barkanyi, T. Active thermal insulation as an element limiting heat loss through external walls. *Energy Build*. 2019, *205*.

Kisilewicz, T.; Fedorczak-Cisak, M.; Sadowska, B.; Ickiewicz, I.; Barkanyi, T.; Bomberg, M.; Gobcewicz, E. On the results of long-term winter testing of active thermal insulation. Energy &. Buildings. 2023, *296* Oct, 2023, 113412.

Klõšeiko, P.; Kalamees, T. Case Study: In-situ Testing and Model Calibration of Interior Insulation Solution for an Office Building in Cold Climate. CESB (Central Europe Symp. On Building), 2016, 2016.

Klõšeiko, P.; Kalamees, T. Long term measurements and HAM modelling of an interior insulation solution for office building in cold climate. In 7th International Building Physics Conference (IBPC2018); Syracuse, NY, USA; Sept. 23–23, 2018, 2018, Syracuse, U, pp. 1423–1428.

Kosuke, S.; Kataoka, E.; Horikawa, S. Thermo-active building system creates comfort, energy efficiency. J. ASHRAE. March 2020, *62*(3), 42–50. ASHRAE.org.

Kuhn, T.S. The Structure of Scientific Revolution. The Chicago U. Press: IL, USA, 1970, see also The Fourth Industrial Revolution | Essay by Klaus Schwab | Britannica, accessed dec 17, 2023.

Lingo, L. Jr.; Roy, U. Novel use of geo solar exergy and storage technology in existing housing applications: Conceptual study. *J. Energy Eng*. 2016, *143*, 0401602.

Meadows, D. The limits of growth, The Donella Meadows Project, Academy of System Change, on the internet; see also "Meadows on social paradigms Meadows. D.H.; Meadows, D.L.; Randers, J.; Behrens III, W.W., The Limits to Growth, Universe Books: New York, NY, USA, 1972.

R2000 standards. https://natural-resources.canada.ca/energy-efficiency/homes/professional-opportunities
/r-2000-standard-for-builders/20564.

Romanska-Zapala, A.; Bomberg, M.; Fedorczak-Cisak, M.; Furtak, M.; Yarbrough, D.; Dechnik, M. Buildings
with Environmental Quality Management (EQM),2018, part 2: Integration of hydronic heating/cooling
with thermal mass. *J. Build. Phys.* 2018, *41*, 397–417.

Romanska-Zapala, A.; Bomberg, M.; Yarbrough, D. Buildings with Environmental Quality Management
(EQM), part 4: A path to the future NZEB. *J. Build. Phys.* 2018a 2018, *43*, 3–21.

Romańska-Zapała, A.; Furtak, M.; Fedorczak-Cisak, M.; Dechnik, M. The Need for Automatic Bypass Control
to Improve the Energy Efficiency of a Building Through the Cooperation of a Horizontal Ground Heat
Exchanger with a Ventilation Unit During Transitional Seasons: A Case Study. In WMCAUS 2018,
Prague IOP Conference Series: Materials Science and Engineering, Vol. 246, 2018b.

Romanska-Zapala, A.; Bomberg, M. Can artificial neuron networks be used for control of HVAC in
environmental quality management systems? In Proceedings of the Central European Symposium of
Building Physics, Prague, Czech Republic, 23–26 September 2019.

Thorsell, T.; Bomberg, M. Integrated methodology for evaluation of energy performance of the building
enclosures. P3: Uncertainty in thermal measurements. *J. Build. Phys.* 2011, *35*, 83–96.

Torrie, R.; Bak, C. Building Back Better with a green renovation wave, (Planning for a green recovery).
*Internet Newsl.* 2022 April 22, 2022, (own archives, accessed on 25 February 2023).

Yarbrough, D.; Bomberg, M.; Romanska-Zapala, A. On the next generation of low energy buildings. *Adv.
Build. Energy Res.* 2019, DOI /10.1080/ 17512549.2019.1692070.

Yarbrough, D.; Bomberg, M.; Romanska-Zapala, A. Buildings with Environmental Quality Management
(EQM), part 3: From log houses to zero-energy buildings. *J. Build. Phys.* 2018, *43*.

Yarbrough David, W.; Bomberg, M.; Romanska-Zapala, A. On the next generation of low energy buildings.
*Adv. Build. Energy Res.* 2021 2021, *15*, A Paradigm Shift in Integrated Building Design -Towards
Dynamically Operated Buildings. https://doi.org/10.1080/17512549.2019.1692070.

# Chapter 8
# Closure: who could start scientific revolution in retrofitting?

The history of residential building construction showed that as long as architects or designers were able to select materials from ever-expanding material developments, the tradition was well preserved and, with local differences, it served the whole society well. Yet, after WWII, when social forces demanded a faster rate of building and material selection grew from 500 in the 1930s to 55,000 in the late 1950s, the use of tradition reached a plateau. New values, based on the speed of construction, cheaper and better materials, started modifying the construction objectives. Finally, in the 1980s, sustainability concepts brought ecological considerations, and socio-economic concepts replaced the traditional wisdom. We must highlight that North America (NA) woke up first, and, observing the failure of the market to follow the best technology, NA instituted national public-private concepts to support Canadian and Scandinavian thinking on building systems. Japan followed soon.

Some philosophers believe the progress comes only from conflicts, and reviewing building technology, we found a conflict between new buildings and retrofitting of existing ones because, in the last three decades, we have only seen fragmentary research without any coordination. We claim that, since 2020, a new construction technology delivers a triple win: for occupants, the economy, and society, but this does not include retrofitting. We reported examples from Canada and the USA that changed the market place in North America, and we claim that without creating similar national public-private programs, reaching a sustainable built environment may take another three decades. Below are our closing comments.

## 8.1 The logic of newly proposed technology

Why are the best, demonstrated-in-practice energy solutions not followed? We believe the reason is that the market pull was missing, that is, there were no visible social needs for this achievement. In other words, the people who buy houses or dwellings are not the same as the people who talk about the need to save energy. So, why is the talk about saving energy not coming to shape the market forces? Our answer is not "politically correct" – we think that there are no more intelligent media to convey the best technical messages. Social media replaced newspapers, radio, or even TV, are today self-centered. McLuhan once said that TV is a medium which replaced the message.

In this situation, a small group of scientists from different countries worked on and off for 20 years and assembled the best of documented advances. Our progress was

https://doi.org/10.1515/9783112217023-008

documented in a number of papers, which at the end created passive and thermo-active cluster (PTAC) technology. Furthermore, we included the basic knowledge on heat, air, water, and vapor transfers, and testing to create a next generation of building physics.

We propose an approach called PTAC technology because it expands the impact of the passive house approach. PTAC is a universal, climate-adapted, public domain technology developed with a view to become a logistic umbrella, in which different options, for example, artificial neural networks, can interact with commercial service and comfort systems. Furthermore, we noticed that designing a cluster of buildings is more effective than a single building because it enables placing in-ground water tanks for the water-sourced heat pump. Therefore, our definition of a cluster also involves a building with surrounding ground. Otherwise, these tanks can be placed on bathroom walls, in staircases or basements, or even in a shed located outside the building.

We do not propose any specific proprietary technology, but rather a blueprint for many different options. It is for a designer to select the water tank capacity, the power of the water source heat pump (WSHP) that operates in the night only, or accept extended hours of WSHP operation, use a one- or two-step water buffer system, and determine the means to ensure the minimum temperature of the low HP terminal. If the supply system is too expensive, the designer may increase the level of thermal insulation in the panels which are mounted on existing walls, or introduce phase change materials, use reflective surfaces in the heating/cooling panels, or increase the area of retrofitting to cover all interior partitions.

The purpose of PTAC technology is to provide a new tradition that expands the possibilities of trade-offs used for the generation, storage, and conservation of thermal energy. Currently, thermal storage is strongly underutilized, and more demonstration work in this area is needed. While photovoltaics may be used in new construction, in retrofitting they are not readily available, and improving the ways of energy supply is necessary. Furthermore, we stress the need for dealing with thermal energy because the transition to electrical storage often may not be economical.

## 8.2 The critical issue

The authors suggested: A revolution like that of Henry Ford in car manufacturing is needed in the retrofitting of existing buildings to provide a win-win-win for society, the economy, and the building's occupants. Society wins with slowing climate change, the economy with plenty of local jobs, and occupants with affordable, good indoor environments. As builders do only what society wants them to do, society should demand buildings with zero carbon emissions and occupants' higher comfort of living. Energy will be saved automatically.

The authors reported a failure of the 1978 demonstration passive house because builders improved the design to reduce the cost but destroyed the indoor environment. They also gave an example of a district between four streets in Montreal, built in a multistage construction process and self-financed by the occupants. These were affordable, low-rise, energy-efficient multiunit residential buildings with 92% energy savings versus requirements of the 2004 building energy code. It had not been followed because retrofitting is not a social issue.

To capture public sentiment, one must create a public-private national retrofitting program. Such a program could be based on the experience from Canadian R2000 standards (1985–2000) and American Building America (1990–2010) and developed simultaneously from public and private ends. Public education on the role of buildings changing from energy user to energy supplier, and privately run retrofitting demonstrations that are 50/50 private/public funded, would be an excellent basis of such a project. The single most important element of the past North American Programs was the vision of the program. There is a need for vision in retrofitting and consensus that public acceptance is needed for the scientific revolution, and this, in turn, depends on the degree of active public participation in the project.

While slowing climate change is a necessity, the current government's grant policy for green products does not make an impact. The public-private demonstration projects are needed to start the social wave on retrofitting with a view to slow climate change. This time around, however, the building physics community must become an active contributor, because the next-generation technology will be based on advanced science.

We also highlighted that PTAC technology is only the starting point for a national, public-private consortium combining public education and private technology demonstrations in one socio-economic package. Our experience from Canada and the USA indicates that without creating a scientific revolution linking retrofitting with climate change solution, we cannot address the failure of retrofitting.

# References

Bomberg, M.; Pazera, M. Methods to check reliability of material characteristics for use of models in real time hygrothermal analysis. In Res. in Building Physics – Proc.1st Central Eur. Symp. Building Physics (eds Gawin and Kisielewicz); Cracow–Lodz: Poland, 2010, 13–15 September 2010, pp. 89–107.

Bomberg, M.; Onysko, D., (Eds.). Energy Efficiency and Durability of Buildings at the Crossroads. 2008. 2008. http://thebestconference.org/BEST1 (accessed on 25 February 2020).

Building America. 1990, https://www.energy.gov/eere/buildings/articles/building-america-program

Buratti, C.; Vergoni, M.; Palladino, D. Thermal comfort evaluation within non-residential environments: development of Artificial Neural Network by using the adaptive approach data, 6th Int. Building Physics Conf., IBPC 2015. *Energy Procedia*. 2015, *78*, 2875–2880.

De Deer, R.; Zhang, F. Dynamic environnent, adaptive comfort, and cognitive performance. In Proceedings of the 7th International Building Physics Conference, IBPC2018, Syracuse, NY, USA, 23–26 Sept. 2018; pp. 1–6

Ferreira, P.; Silva, S.; Ruano, A.; Negrier, A.; Conceição, E. Neural Network PMV Estimation for Model-Based Predictive Control of HVAC Systems. In WCCI 2012 IEEE World Congress on Comp. Intelligence, June 10–15, 2012 2012, Brisbane, Australia, pp. 15–22, doi: 10.1109/IJCNN.2012.6252365666

Fort, J.; Kocí, J.; Pokorný,; Podolka, L.; Kraus, M.; Cerný, R. Characterization of Responsive Plasters for Passive Moisture and Temperature Control. *Appl. Sci.* 2020, *10*, 9116, http://dx.doi.org/10.3390/app1024911.

Häupl, P.; Grunewald, J.; Fechner, H. Moisture behavior of a "Gründerzeit" -house by means of a capillary active interior insulation. In Proceedings of the Building Physics in the Nordic Countries, Gothenburg, Sweden, 24–26 August 1999, pp. 225–232 .

Lingo, L. Jr.; Roy, U. Novel Use of Geo solar Exergy and Storage Technology in Existing Housing Applications: Conceptual Study. *J. Energy Eng.* 2016, *143*, 0401602.

Sato, K.; Kataoka, E.; Horikawa, S. Thermo-Active Building System Creates Comfort, Energy Efficiency. J. ASHRAE. March 2020, *62*(3), 42–50. ASHRAE.org.

Meadows, D. The limits of growth, The Donella Meadows project, Academy of System Change, on the internet; see also "Meadows on social paradigms Meadows, D.H.; Meadows, D.L.; Randers, J.; Behrens III, W.W. The Limits to Growth. Universe Books: New York, NY, USA, 1972.

Torrie, R.; Bak, C. Building Back Better with a green renovation wave, (Planning for a green recovery). *Internet Newsl.* 2022 April 22, 2022, (own archives).

Simonson, C.J.; Salonvaara, M.; Ojanen, T. Heat and mass transfer between indoor air and a permeable and hygroscopic building envelope: P. II, Verification, and numerical studies. *J. Bldg. Phys.* 2004 2004, *28*, 161–185.

Vereecken, E.; Roels, S. Capillary active interior insulation: do the advantages really offset potential disadvantages? *Mater. Struct.* 2015 2015, *48*(9), 3009–3021.

# Further reading: historic sources of information

Acone, M.; Romano, R.; Piccolo, A.; Siano, P.; Loia, F.; Ippolito, M.G.; Zizzo, G. Designing an Energy Management System for smart houses. In IEEE 15th International Conference on Environment and Electrical Engineering (EEEIC), 2015, 2015, pp. 1677–1682.

Bomberg, M.T.; Kumaran, M.K. A Test Method to Determine Air Flow Resistance of Exterior Membranes and Sheathings. *J. Build. Phys.* 1986, *9*, 224–235.

Bomberg, M.; Wojcik, R.; Piotrowski, J. A concept of integrated environmental approach, Part 2: Integrated approach to rehabilitation. *J. Build. Phys.* 2016, *39*(6).

Brennan, T.H.; Stack, H.K.; Bomberg, M. Quality assurance and commissioning process in High Env. Performance (HEP) demonstration house in NY State. 2008, www.thebestconference.org/best1.

Ciafrone, C.; Roppel, P.; Hardock, D. Holistic approach to achieving low energy high-rise residential buildings. *J. Build. Phys.* 2016, *39*(5). At: https://www.brikbase.org/sites/default/files/BEST4_2.2cian frone.pdf.

CHBA Canadian Home Builders' Association Builders' Manual, R-2000 Project, 1989.

CMHC. Ventilation and Airtightness of New Detached Canadian Housing. 1990, Research Div.

CMHC Field Investigation Survey of Airtightness, Air Movement, and Indoor Air Quality in High Rise Apartment Buildings, Summary Report for CMHC, 1993.

CMHC. 1995. EASE Demonstration House APCHQ's Advanced House.

CMHC. Field Investigation of Indoor Environment and Energy Usage in Mid-Rise Residential Buildings. 1997.

Clements-Croome, D. et al. Intelligent buildings: design, management and operation. Thomas Telford Publishing: London, 2004.

Demarais, G.; Derome, D.; Fazio, P. Mapping of air leakage in exterior walls. *J. Build. Phys.* 2000, *24*, 132–154.

Dickens, H.B.; Hutcheon, N.B. Moisture Accumulation in Roof Spaces Under Extreme Winter Conditions. RILEM/CIB Symp. Moisture Problems in Buildings. 1965, 1.

Duszczyk, K. Inteligentny budynek – nowoczesne technologie w laboratorium dydaktycznym. *Przeglad Elektrotechniczny.* 2006, nr 10, 6–8.

Eyre, D. Air-Vapor Barrier Manual, Sask. Res. Council, Sask., SK, 1981.

Finch, G.; Burnett, E.; Knowles, W. Energy Consumption in Mid- and High-Rise Residential Buildings in British Columbia, EE3–3, 2010. Available at: www.thebestconcerence. org/BEST2.

Forest, T. Drying of Walls Prairie Region, 1990.

Ganguli, U. Wind and Air Pressures on the Building Enclosure, an Air Barrier for the Building Enclosure, NRC. *Bldg. Sci. Insight.* 1986, 7–13.

Garden, G.K. Control of Air Leakage is Important CBD 72. NRC: DBR, 1965.

Greig, A.R. Wall Insulation, U. of Saskatchewan, College of Engineering. Bul. No. 1. 1922.

Handegord, G.O. Vapor Barriers in Home Construction CBD 9. NRC: DBR, 1960.

Hechler, F.G.; McLaughlin, E.R.; Queer, E.R. Simultaneous Heat and Vapor Transfer Characteristics of an Insulating Material. *ASHVE Trans.* 1942, *48*, 505.

Hurtado, L.A.; Nguyen, P.H.; Kling, W.L.; Zeiler, W. Building Energy Management Systems – Optimization of comfort and energy use, Power Engi. Conference (UPEC), 48th International Universities', 2013, pp. 1–6.

Hutcheon, N.B. The Utility of Building Science, Reprint of the Lecture Delivered in 1971. *J. Thermal Env. and Bldg. Sci.* 1998, *22*, 4–9.

Hutcheon, N.B.; Handegord, G.O. (1980). Evolution of the Insulated Wood-Frame Wall in Canada. In: Proceedings of the 8th CIB Congress, Oslo. Vol. 16, pp. 434–438.

Jabłoński, A. Zadania integracji systemów w budynkach inteligentnych. Przeglad Elektrotechniczny. 2008, nr 7, 182–185.

https://doi.org/10.1515/9783112217023-009

Joy, F.A.; Queer, E.R.; Schreiner, R.E. (1948). Water Vapor Transfer through Building Materials: Bulletin No. 61, Pennsylvania State College, Engineering Experiment Station.

Kent, A.D.; Handegord, G.O.; Robson, D.R. A Study of Humidity Variations in Canadian Houses. *ASHRAE Trans.* 1966, *72*, Part II, 11.1.1–11.1.8.

Kwasowski, P.; Fedorczak-Cisak, M. Wpływ zintegrowanych systemów automatyki na efektywność energetyczną budynków w świetle normy PN–EN 15232, Fizyka budowli w teorii i praktyce, pod red. P. Klemm, D. Heim, K. Klemm, M. Wojtczak, Instytut Fizyki Budowli Katarzyna i Piotr Klemm S.C., Łódź, 2013, 53–58.

Lstiburek, J.W.; Lischkoff, J.K. A New Approach to Affordable, Low Energy House Construction. 1984, Alberta Department of Housing.

Lstiburek, J.W. Two Case Studies with Failures in the Environmental Control of Buildings. *J. Build. Phys.* 1995, *19*, 149–172.

Lstiburek, J.W.; Pressnail, K.; Timusk, J. Transient Interaction of Buildings with HVAC Systems – Updating the state of the art. *J. Thermal Env. Build. Sci.* 2000, *24*, 111–131.

Lux, M.E.; Brown, W.C. Air Leakage Control, an Air Barrier for the Building Enclosure. NRC Bldg. Sci. Insight. 1986, 13–19.

Łukaszewski, R.; Winiecki, W. Systemy do monitorowania zużycia energii elektrycznej w domu. Przeglad Elektrotechniczny. 2014, nr 11, 35–38.

Manic, M.; Wijayasekara, D.; Amarasinghe, K.; Rodriguez-Andina, J.J. Building Energy Management Systems: The Age of Intelligent and Adaptive Buildings. *IEEE Ind Electron Mag.* 2016. *10*(Iss. 1), 25–39.

MH. Structural Requirements for Air Barriers and Testing of Air Barrier Construction Details. 1991, Morrison Hershfield Ltd.

Mikulik, J. et al. Inteligentne budynki – informacja i bezpieczeństwo, Kraków, Wydawnictwo LIBRON. 2016.

Muszyński, L. Inteligentne pomiary zużycia mediów budynku w systemie KNX. Przeglad Elektrotechniczny. 2010, nr 4, 302–304.

Niezabitowska, E. et al. Budynek inteligentny, Tom I Potrzeby użytkownika, a standard budynku inteligentnego, Wydawnictwo Politechniki Śląskiej. Gliwice, 2010.

Niezabitowska, E.; Mikulik, J. Budynek inteligentny, Tom II Podstawowe systemy bezpieczeństwa w budynkach inteligentnych, Wydawnictwo Politechniki Śląskiej. Gliwice, 2014.

Noga, M.; Ożadowicz, A.; Grela, J. Efektywność energetyczna i Smart Metering – nowe wyzwania dla systemów automatyki budynkowej. Napedy i Sterowanie. 2012, nr 12, 54–59.

NRC. Testing Air Barrier Systems for Wood Frame Walls. 1988, IRC/NRC.

NRC. Establishing the Protocol for Measuring Air Leakage and Air Flow Patterns in High-Rise Apartment Buildings. 1990, Institute for Research in Construction, NRC.

NRC Building Performance News. Issue No. 17, 1993.

Orr, H.W. Condensation in Electrically Heated Houses. In 2nd International CIB/RILEM Symposium on Moisture Problems in Buildings, 1974.

Ożadowicz, A. Analiza porównawcza dwóch systemów sterowania inteligentnym budynkiem: systemu europejskiego EIB/KNX oraz standardu amerykańskiego na bazie technologii Lon Works, rozprawa doktorska, promotor: Zbigniew Hanzelka. AGH, Kraków, 2007.

Ożadowicz, A. Automatyka budynkowa w realizacji systemów smart grid – energooszczędność i integracja na poziomie odbiorcy energii. *Wiadomosci Elektrotechniczne.* 2013, nr 11, 38–42.

Ożadowicz, A. Systemy automatyki budynkowej jako element inteligentnych sieci elektroenergetycznych – Smart Mering i aktywny odbiorca. *Napedy i Sterowanie.* 2013, nr 12, 52–55.

Ożadowicz, A.; Grela, J., Aktywni odbiorcy i standardy automatyki budynkowej jako element Smart Meteringu w budynkach, http://eip-online.pl/, 2015.

Pamuła, A.; Papińska-Kacperek, J. Inteligentne domy i i inteligentne sieci energetyczne jako element infrastruktury Smart City, Zeszyty Naukowe Uniwersytetu Szczecińskiego. Studia Informatica. 2012, nr 29, 57–69.

Perrault, J.C. Air Barrier Systems: Construction Applications, An Air Barrier for the Building Enclosure. *NRC Bldg. Sci. Insight*. 1986, 20–24.

Quirouette, R.L. The Difference Between a Vapor Barrier and an Air Barrier. Building Practice Note 54. NRC: IRC, 1985.

Quirouette, R.L. The Air Barrier Defined, an Air Barrier for the Building Enclosure. *NRC Bldg. Sci. Insight*. 1986, 1–7.

Romańska-Zapała, A. Model badawczy zintegrowanego systemu sterowania procesami na przykładzie budynku Małopolskiego Laboratorium Budownictwa energooszczędnego. *Materialy Budowlane*. 2014, *12*, 31–33.

Romańska-Zapała, A. Zintegrowane systemy sterowania procesami w obiektach budowlanych. *Materialy Budowlane*. 2014, *5*, 115–116.

Rowley, F.B.; Algren, A.B.; Lund, C.E. Condensation Within Walls, and Condensation and Moisture in Relation to Building Construction and Operation. *ASHVE Trans*. 1938, *44*.

Rowley, F.B. A Theory Covering the Transfer of Vapor through Materials. *ASHVE Trans*. 1939, *45*, 545.

Sasaki, J.R.; Wilson, A.G. Window Air Leakage CBD 25. NRC: DBR, 1962.

Sasaki, J.R.; Wilson, A.G. Air Leakage for Residential Windows. *ASHRAE Trans*. 1965, *71*(Part 11), 81–88.

Saha, A.; Kuzlu, M.; Khamphanchai, W.; Pipattanasomporn, M.; Rahman, S.; Elma, O.; Selamogullari, U.S.; Uzunoglu, M.; Yagcitekin, B. A home energy management algorithm in a smart house integrated with renewable energy. In IEEE PES Innovative Smart Grid Technologies, Europe, 2.

Shao, H.; Fu, H. Design and Implementation of Intelligent Building Engineering Information Management System. In Intelligent Computation Technology and Automation (ICICTA), 7th International Conference on, 2014, 2014, pp. 158–161.

Sherman, M.H.; Chan, R. Research and Practice Lawrence Berkeley Nat Lab Report No LBNL-53356, 2004, 2004.

Sroczan, E. Algorytmy projektowania instalacji w budynkach inteligentnych. *Wiadomosci Elektrotechniczne*. 2007 nr 5, 24–30014, 1–6.

Sroczan, E. Funkcje zintegrowanych systemów zarządzania energią w obiektach inteligentnych. *Wiadomosci Elektrotechniczne*. 2009, nr 12, 35–39.

Stricker, S. Measurement of Airtightness of Houses. *ASHRAE Trans*. 1975, 81(1), 148–167.

Sun, B.; Luh, P.B.; Qing-Shan, J.; Jiang, Z.; Wang, F.; Song, C. Building Energy Management: Integrated Control of Active and Passive Heating, Cooling, Lighting, Shading, and Ventilation Systems. *IEEE Trans. Autom. Sci. Eng*. 2013, *10*(Iss. 3), 588–602.

Tamura, G.T.; Wilson, A.G. Air Leakage and Pressure Measurements on Two Occupied Houses. *ASHRAE J*. 1963, *5*(12), 65–73.

Tamura, G.T.; Wilson, A.G. Pressure Differences for a Nine-Storey Building. *ASHRAE Trans*. 1966, *72*(Part 1), 180–189.

Tamura, G.T.; Wilson, A.G. Building Pressures Caused by Chimney Action, and Pressure Differences Caused by Chimney Effect in Three High Buildings. *ASHRAE Trans*. 1967, *73*, Part II.

Tamura, G.T.; Kuester, G.H.; Handegord, G.O. Condensation Problems in Flat Wood-Frame Roofs. In 2nd Int. CIB/RILEM Symp. on Moisture Problems in Buildings, 1974.

Tamura, G.T. Measurement of Air Leakage Characteristics of House Enclosures. *ASHRAE Trans*. 1975, 81(1), 202–211.

Teesdale, L.V. Comparative Resistance to Vapor Transmission of Various Building Materials. *ASHVE Trans*. 1943, 49, 24.

Torpe, G.; Graee, T. The Effects of Air Currents in Moisture Migration and Condensation in Wood- Wang S., Intelligent Buildings and Building Automation, Taylor & Francis, 2009Frame Structures. Bygg. 1961, 9 (l), 1–10. Available as NRCC TT-1478.

Trow. Criteria for Air Leakage Characteristics of Building Enclosures – Final Report, 1989.

Timusk, J. Moisture Control in Wood Framed Wall Assemblies. In 6th Conference on Building Science and Technology, 1992.

Vergini, S.E.; Groumpos, P.P. A review on Zero Energy Buildings and Intelligent Systems, Information, Intelligence, Systems and Applications (IISA). In 2015 6th International Conference on, 6–8 July 2015.

Wang, S. Intelligent Buildings and Building Automation. Taylor & Francis, 2009.

Walker, J.S.; Wilson, D.J.; Sherman, M.H. A comparison of power law to quadratic formulations for air infiltration calculation. *Energy Build*. 1997, *27*(3), June 1993.

Wicaksono, H.; Rogalski, S.; Kusnady, E. Knowledge-based intelligent energy management using building automation system. In IPEC, 2010 Conference Proceedings, 27–29 Oct. 2010.

Wilson, A.G.; Nowak, E. Condensation Between Panes of Double Windows. *ASHRAE Trans*. 1959, 65, 551–570.

Wilson, A.G.; Solvason, K.R.; Nowak, E.S. Evaluation of Factory-Sealed Double-Glazed Window Units. ASTM, STP 251. 1959, 3–16.

Wilson, A.G. Influence of the House On Chimney Draft. *ASHRAE J*. 1960a, 2(12), 63.

Wilson, A.G. Condensation on Inside Window Surfaces. CBD 4. NRC: DBR, 1960b.

Wilson, A.G. Air Leakage in Buildings. CBD 23. NRC: DBR, 1961.

Wilson, A.G.; Garden, G.K. Moisture Accumulation in Walls Due to Air Leakage. In: RILEM/CIB Symposium on Moisture Problems in Buildings, Vol. 1. Helsinki, 1965.

Wilson, A.G. Moisture in Canadian Wood-Frame Housing Construction: Problem, Research and Practice from 1975 to 1991. Report to CMHC, 1992.

# About the authors

The authors are overachievers. On average, the three individuals hold 4.7 academic titles (diplomas), work across 1.7 continents, and have 46 years of professional experience and 250 published papers each. From this perspective, they decided to write a white paper on the improvement path for building energy and the indoor environment.

**Mark Bomberg** (D.Sc. (Eng), Warsaw, Poland), Tech.D. at Lund University, Sweden, is presently a research professor of mechanical engineering at Clarkson University in Potsdam, NY, USA. He worked for 34 years at the National Research Council of Canada and has taught university courses in building science/physics/ civil or mechanical engineering in the USA, Canada, Germany (guest), Poland, and China. He has about 100,000 recorded readers and is listed at second place worldwide in the specialization of thermal insulation by Research Gate in Berlin.

**David W. Yarbrough** retired from Tennessee Technological University in 2002 as professor and chair of chemical engineering. He retired from the research staff at the Oak Ridge National Laboratory in 2011. He is an active member and fellow of the American Society for Testing and Materials. He is a graduate of Georgia Tech (B.Ch.E. 1960, M.S. 1961, Ph.D. 1966). He has also been recognized with awards from Tennessee Tech, the Tennessee Academy of Science, and the International Thermal Conductivity Conference. He is the author or co-author of more than 250 papers and presentations. He is presently employed by R&D Services, Inc., a company he co-founded in 1994.

**Hamed H. Saber** is a professor in the Mechanical Engineering Department at Jubail Industrial College and the chair of the Research and Consultations Department at the Deanship of Research and Industrial Development, Royal Commission for Jubail and Yanbu, Saudi Arabia. He previously held research and academic positions at the Institute for Space and Nuclear Studies and the Department of Chemical and Nuclear Engineering at the University of New Mexico (USA), as well as at the National Research Council of Canada. He earned his Ph.D. from the University of New Mexico in 2000 and holds B.Sc. (1988) and M.Sc. (1992) from Mansoura University, Egypt. He has received several awards, including the "International Alexander Schwartz Award," and is listed among the top 2% of scholars worldwide, according to a study conducted by Stanford University.

https://doi.org/10.1515/9783112217023-010

# Index

https://doi.org/10.1515/9783112217023-011

www.ingramcontent.com/pod-product-compliance
Lightning Source LLC
Chambersburg PA
CBHW081530220326
41598CB00036B/6387